Recent Titles in This Series

120 **Takeyuki Hida and Masuyuki Hitsuda,** Gaussian processes, 1993
119 **M. V. Karasev and V. P. Maslov,** Nonlinear Poisson brackets. Geometry and quantization, 1993
118 **Kenkichi Iwasawa,** Algebraic functions, 1993
117 **Boris Zilber,** Uncountably categorical theories, 1993
116 **G. M. Fel′dman,** Arithmetic of probability distributions, and characterization problems on abelian groups, 1993
115 **Nikolai V. Ivanov,** Subgroups of Teichmüller modular groups, 1992
114 **Seizô Itô,** Diffusion equations, 1992
113 **Michail Zhitomirskiĭ,** Typical singularities of differential 1-forms and Pfaffian equations, 1992
112 **S. A. Lomov,** Introduction to the general theory of singular perturbations, 1992
111 **Simon Gindikin,** Tube domains and the Cauchy problem, 1992
110 **B. V. Shabat,** Introduction to complex analysis Part II. Functions of several variables, 1992
109 **Isao Miyadera,** Nonlinear semigroups, 1992
108 **Takeo Yokonuma,** Tensor spaces and exterior algebra, 1992
107 **B. M. Makarov, M. G. Goluzina, A. A. Lodkin, and A. N. Podkorytov,** Selected problems in real analysis, 1992
106 **G.-C. Wen,** Conformal mappings and boundary value problems, 1992
105 **D. R. Yafaev,** Mathematical scattering theory: General theory, 1992
104 **R. L. Dobrushin, R. Kotecký, and S. Shlosman,** Wulff construction: A global shape from local interaction, 1992
103 **A. K. Tsikh,** Multidimensional residues and their applications, 1992
102 **A. M. Il′in,** Matching of asymptotic expansions of solutions of boundary value problems, 1992
101 **Zhang Zhi-fen, Ding Tong-ren, Huang Wen-zao, and Dong Zhen-xi,** Qualitative theory of differential equations, 1992
100 **V. L. Popov,** Groups, generators, syzygies, and orbits in invariant theory, 1992
99 **Norio Shimakura,** Partial differential operators of elliptic type, 1992
98 **V. A. Vassiliev,** Complements of discriminants of smooth maps: Topology and applications, 1992
97 **Itiro Tamura,** Topology of foliations: An introduction, 1992
96 **A. I. Markushevich,** Introduction to the classical theory of Abelian functions, 1992
95 **Guangchang Dong,** Nonlinear partial differential equations of second order, 1991
94 **Yu. S. Il′yashenko,** Finiteness theorems for limit cycles, 1991
93 **A. T. Fomenko and A. A. Tuzhilin,** Elements of the geometry and topology of minimal surfaces in three-dimensional space, 1991
92 **E. M. Nikishin and V. N. Sorokin,** Rational approximations and orthogonality, 1991
91 **Mamoru Mimura and Hirosi Toda,** Topology of Lie groups, I and II, 1991
90 **S. L. Sobolev,** Some applications of functional analysis in mathematical physics, third edition, 1991
89 **Valeriĭ V. Kozlov and Dmitriĭ V. Treshchëv,** Billiards: A genetic introduction to the dynamics of systems with impacts, 1991
88 **A. G. Khovanskiĭ,** Fewnomials, 1991
87 **Aleksandr Robertovich Kemer,** Ideals of identities of associative algebras, 1991
86 **V. M. Kadets and M. I. Kadets,** Rearrangements of series in Banach spaces, 1991

(Continued in the back of this publication)

Translations of

MATHEMATICAL MONOGRAPHS

Volume 120

Gaussian Processes

Takeyuki Hida
Masuyuki Hitsuda

Translated by
Takeyuki Hida and Masuyuki Hitsuda

American Mathematical Society
Providence, Rhode Island

ガウス過程

表現と応用

GAUSU KATEI (Gaussian Processes)

by Takeyuki Hida and Masuyuki Hitsuda

Originally published in Japanese by Kinokuniya Company Ltd., Publishers, Tokyo in 1976

Translated from the Japanese by Takeyuki Hida and Masuyuki Hitsuda

2000 *Mathematics Subject Classification*. Primary 60G15; Secondary 60J99.

Abstract. This book provides a general theory of representation of Gaussian processes by using Brownian motion. The book is unique in several ways: (1) it describes the basic idea used to investigate the probabilistic structure of Gaussian processes in terms of the canonical representation; (2) it is a self-contained monograph involving basic notions and application; (3) multiple Markov properties of a Gaussian process and equivalence problems are discussed as important developments.

The Library of Congress has cataloged an earlier hardcover edition as follows:

Hida, Takeyuki, 1927–

[Gausu katei. English]

Gaussian processes/Takeyuki Hida and Masuyuki Hitsuda; translated by Takeyuki Hida and Masuyuki Hitsuda.

p. cm. — (Translations of mathematical monographs, ISSN 0065-9282; v. 120)

Translation of: Gausu katei.

Includes bibliographical references and index.

ISBN 0-8218-4568-3 (acid-free paper)

1. Gaussian processes. I. Hitsuda, Masuyuki, 1938– . II. Title. III. Series.

QA274.4.H5313 1993 92-32176

519.2—dc20 CIP

AMS softcover ISBN 978-0-8218-4358-1

Reprinted by the American Mathematical Society, 2007.
Translation authorized by the Kinokuniya Company Ltd.

Printed in the United States of America.

Information on copying and reprinting can be found in the back of this volume.
∞ The paper used in this book is acid-free and falls within the guidelines established to ensure permanence and durability.
Visit the AMS home page at http://www.ams.org/

10 9 8 7 6 5 4 3 2 1 12 11 10 09 08 07

Contents

Preface to the English Edition

We are pleased to have the English edition of our book "Gaussian Processes" published by the American Mathematical Society. Through this translation, our book will reach a more international audience, and canonical representation theory will be more widely used, to our great joy.

We want to take this opportunity to explain the significance and importance of the representation theory of Gaussian processes. This theory was initiated in 1955 by Paul Lévy, who emphasized its important roles in stochastic process theory. Since then, numerous developments have been made by many authors. Still, there can often be serious misunderstandings about the role of the theory. For instance, someone might think that, for a given Gaussian process, its covariance function (and the mean) tell us everything, and so there is no need for anything else. While the covariance function certainly gives much information about the probabilistic structure of the given Gaussian process, it does not yield finer visualized structures such as properties of sample functions, formulas for the best predictor of future values, multiple Markov properties, and so forth.

Once the canonical representation is obtained, these properties can be easily derived from those of the Brownian motion and the analytic expression of the kernel function. Thus, readers will find in our book many advantages of the canonical representation theory.

We are also happy to have been given the opportunity to add to the original edition some new results on representation theory.

Acknowledgements. The authors are grateful to Professor I. Kubo of Hiroshima University for the use of his software "Word Processor for Mathematical Papers", which was helpful in writing this manuscript. It is also our pleasant duty to record our gratitude to Drs. Si Si (University of Yangon) and K.-S. Lee (Korea University) who read most of the chapters and gave us their comments and corrections.

Takeyuki Hida
Masuyuki Hitsuda
November 1991

Preface

Two years have passed since we began to write this book "Gaussian Processes". One of our main aims is to popularize the beautiful theory of Gaussian processes, not only to mathematicians, but also to physicists and engineers as well as others who use mathematics in their work. The readers can, we believe, see our ideas in this book, although it is far from what we originally intended.

Gaussian processes, being mathematical models of random phenomena that are Gaussian in distribution, are very interesting research objects in probability theory, and they are quite useful in applications. In this book we emphasize the importance of canonical representation theory. This theory, which was originated by P. Lévy in 1955, has not yet been well popularized. We believe that representation theory always appears as a basic tool in theories concerned with *causality* where the development over time is explicitly taken into account. This book illustrates this principle.

We start with elementary background material so that the book is accessible to beginners. The topics are arranged to form a systematic introduction to the representation theory for Gaussian processes. The reader will find in the explanatory Notes at the end of the book further developments of this theory. Some of the related subjects, which are somewhat outside of the mainstream of the book, appear in the Appendix.

The celebration of the 200th birthday of C. F. Gauss is coming next year. It is our hope that, on this occasion, the study of the stochastic process bearing his name will be further developed, and its importance be well understood.

The authors are grateful to the editors Mr. K. Uzuoka and Mr. H. Mizuno.

Takeyuki Hida
Masuyuki Hitsuda
February 1976
Nagoya

Introduction

We will consider a mathematical theory of a random phenomenon that evolves in time according to a probabilistic law. This is called a stochastic process and the Gaussian case is of main concern in this book.

The first task is to explain why the Gaussian case is so important. Consider a Gaussian distribution. There are in practice many situations that are subject to a Gaussian distribution, and its mathematical theory has been developed beginning with P. S. Laplace. Even now the importance of this distribution is becoming better recognized and additional meanings are being discovered. The details of this fact are the main concern of this book. Here we present a quick intuitive explanation.

According to Gauss's theory of errors, accumulating small amounts of error in measurement, if they appear independently, yield a distribution that is close to a normal distribution. Such a derivation of a Gaussian distribution is substantial and, in fact, it became a leitmotiv for the mathematical theory of a random event.

We now present a more mathematical description. Let X_n, $n \geq 1$, denote the successive results of observations and assume that the X_n are independent variables. If the same kind of observation is repeated again and again, then the X_n should be subject to the same distribution (not necessarily Gaussian). Denote the mean and the variance by m and σ^2, respectively. Assume that the third order moment is finite. The partial sum of the X_n is denoted by S_n: $S_n = X_1 + X_2 + \cdots + X_n$, and is normalized. That is, we have

$$\frac{S_n - n \cdot m}{\sqrt{n}\sigma}.$$

The probability distribution of the random variable above can be proved to be close to the standard Gaussian distribution. Such a limit theorem has been known empirically for many years. Theorems of this kind are called the *central limit theorem*. They occupy a central position among the limit theorems in probability theory (see Chapter 1, §4). The basic idea of this result has been known for many years, however it can be further investigated and is expected to yield additional applications.

We next describe the statistical properties of a Gaussian distribution. The values are distributed over the entire real line and have density function of the form

$$\frac{1}{\sqrt{2\pi}\sigma} \exp\left[-\frac{1}{2\sigma^2}(x-m)^2\right],$$

where m is the mean and σ^2 is the variance.

We are now in a position to consider characterizations of a Gaussian distribution, through its various properties.

(i) A Gaussian distribution has moments of all orders. The nth order moment around the origin, denoted by μ_n is given by

$$\mu_n = \begin{cases} 0, & \text{for } n \text{ odd}, \\ (n-1)!!\sigma^2, & \text{for } n \text{ even}. \end{cases}$$

The distribution is symmetric around the mean. Both the skewness $\mu_3/\mu_2^{3/2}$ and the kurtosis $\mu_4/\mu_2^2 - 3$ vanish.

(ii) A Gaussian distribution is stable with exponent 2. The family of all Gaussian distributions (or suitable family of them) is closed under the convolution. Note that this operation appears when the distribution of the sum of two independent random variables is computed. As a result, a Gaussian system (see Chapter 2, §1) can be applied to linear operations. Indeed, the linear structure of the system mentioned above is a major characteristic in our investigation. Thus, one should understand the dominant property of the system from this point of view rather than viewing linear operators as easier to deal with.

(iii) To categorize Gaussian distributions simply as distributions that arise only from random phenomena such as errors in some measurements is too narrow. There is much more. For example, consider the uniform probability measure on the n-dimensional sphere with radius $\sqrt{n}$, and take the projection of this measure onto a straight line passing through the origin. The probability measure thus obtained is close to the standard Gaussian distribution if n is large (E. Borel). This observation is familiar in statistical mechanics and other fields. Yet understanding this fact is quite important in analysis. In fact, it may be considered a key property that illustrates connections between several topics in probability theory and those in functional analysis, quantum dynamics, statistical mechanics, and so forth. On the other hand, it indicates how to approximate a Gaussian distribution by distributions over finite intervals.

(iv) Another perspective is given by applications. This perspective comes from historical facts and from proposals for the future. Gaussian distributions have never been considered responsible for all random phenomenon, yet the assumption that a distribution is Gaussian often yields new developments on work coming from applications. Theoretical interpretations will be given soon or later, and in any event it is certain that Gaussian distributions occupy the top position among all probability distributions.

The observations above together along with the characterizations of a Gaussian distribution given in §2.2. will be helpful to those of us who are in search of ways to study Gaussian systems.

We can now discuss a Gaussian process which is a particular Gaussian system with a continuous time parameter. In general, the probability distribution of a Gaussian system is completely determined by its mean vector and the covariance matrix (§2.1). Consider a Gaussian system with mean vector zero. It is inappropriate to think that analytic or algebraic properties of the covariance matrix can immediately tell us all its probabilistic properties. To know more about a Gaussian process it is necessary to investigate further its profound statistical structure. Namely, the passage, from moment to moment, of the random events and the nature of the dependency of the given Gaussian process from past to future should be clarified explicitly. To this end we provide the following powerful methods:

(1) First take the most typical and basic Gaussian process and study its probabilistic properties in detail,

(2) form a representation of a general Gaussian process based on the process determined in (1).

Having agreed with this idea, one can see that the Gaussian process determined by (1) is nothing but a Brownian motion $B(t)$ to be described in Chapter 2. $B(t)$ always has independent increments as time propagates, and the increment is homogeneous both temporally and spatially. There is no other Gaussian process with these properties. $B(t)$ can also be thought of as a continuous analogue of a partial sum of independent identically distributed Gaussian random variables.

We now give a heuristic consideration of Brownian motion. In 1827, Robert Brown discovered an irregular movement of the grains of pollen suspended in the water. Later in 1905, Albert Einstein succeeded in establishing a mathematical theory modeling this phenomenon. Systematic study of this motion as a Gaussian process has been continued by Paul Lévy, Nobert Wiener, and others. Its Markov property, the analytic behavior of the sample paths, the close connections with functional analysis, and so forth, are now well known as far as the one-dimensional parameter case is concerned.

It is Paul Lévy who proposed (1955) the idea stated in (2). Given a Gaussian process $\{X(t); t \geq 0\}$, he wished to find a Brownian motion $\{B(t); t \geq 0\}$ and an integral kernel $F(t, u)$ so that $X(t)$ has a representation of the form

$$X(t) = \int_0^t F(t, u)\, dB(u) \qquad \text{(Wiener integral)}.$$

Such a representation is not unique, of course. Lévy claimed that there exists a canonical représentation the expression of which describes well the properties of the given $X(t)$ in terms of $B(t)$ and $F(t, u)$. The existence and uniqueness of the canonical representation, and how the Markov property and other properties appear in the representation, . . . , these are exactly what we should like to discuss in this book.

Our aim in this book has now become clear. We must, however, acknowledge that applications of our theory are not discussed in detail. What we mean by applications are those subjects that present mathematical problems and stimulate the reader. Mention of a few fields with interesting applications is now in order.

Noise that arises in signal transmissions or in measurements of something is well known as an awkward object disturbing the true values. The noise that appears sequentially in time in such a case is random and often Gaussian in distribution. Idealized noise is white in the sense that it is an equally weighted sum of simple harmonic oscillators with random amplitude. As for the case of telecommunication, the message itself may be considered a stochastic process before we receive the signal. In coding theory, noise is important for evaluating numerically the amount of information gained from the process successively in time. This example can serve to illustrate the notion in Chapter 2 of *innovation.* In cases such as communication and measurement, our interest is in predicting the future values by filtering of the data given up to the present, and thus the idea is similar to canonical representation theory.

Next, consider a biological model. Here we see such examples as spontaneous fluctuation in a primitive animal causing random movement or examples such as stochastic evolutional phenomena in population genetics. These fluctuations, once they are idealized mathematically, can often be considered as Gaussian random functions or else their functionals. We are thus led to describe these random phenomena as functions of Gaussian processes, where time evolution is involved so that Markov properties or stationary states are involved.

Another important field of application deserves mention; namely, quantum mechanics. Since there are so many different ways of connecting Gaussian processes with problems in physics, it is impossible to present them systematically. Incidentally, the role in quantum mechanics played by probability theory might be compared to that played in classical mechanics by differential and integral calculus. Particular examples related to Gaussian processes can be found in field theory in such cases as Euclidean free fields described by stationary Gaussian Markov fields (free fields), where deep properties can be seen by their sample paths; more general Euclidean fields correspond to higher order Markov properties of the Gaussian processes; the T-positivity can be rephrased in terms of a stationary Gaussian random field and so forth. It is important to note that the Gaussian property of the distribution is always assumed. Part of the characteristic behind this assumption will be illustrated in Chapter 5.

Finally we must say that there are two topics, related to the subjects in this book, that we do not include. One is the sample function properties of a Gaussian process, such as sample function continuity or further analytic properties. The second topic is the level crossing number. These topics have been widely developed and we know that they have connections with representation theory. But they have their own technique of study and different purposes. Thus, they are to be discussed within a much wider setup and are far beyond the scope of this book. Still, we will touch upon such facts in Chapter 6, in line with the approach to the equivalence of Gaussian measures.

An additional missed topic is a systematic approach to the canonical representation of Gaussian random fields that have a multidimensional time parameter. As we mentioned earlier, such fields have a close connection to Euclidean fields, so that a general theory is quite attractive for us. To our regret, we are not ready to propose a representation theory. It is our hope to discuss this topic sometime in the near future.

CHAPTER I

Foundations of Probability Theory and Limit Theorems

This chapter serves to introduce some of the basic concepts in probability theory and to illustrate limit theorems, in particular the central limit theorem, from which we are led naturally to Gaussian distributions. We refer to the literature in the respective fields for details of proofs of the theorems, and mention only what we shall need in the following chapters.

§1. Probability spaces and random variables

The foundation of probability theory is formulated mathematically with the help of measure theory and the Lebesgue integral. This approach was proposed by A. N. Kolmogorov in 1933 and since then great success has been made in the limit theorems in probability, the theory of stochastic processes, and related fields.

The notion of probability in mathematics is defined as the measure of a measurable set consisting of random parameters. Our definition must therefore start with a nonempty set Ω of all the random parameters ω in question. A system $\mathbf{B}$ of subsets of Ω is called a *completely additive field* or simply *σ-field*, if the following three conditions are satisfied:

($\mathbf{B}$.1) $\Omega \in \mathbf{B}$,
($\mathbf{B}$.2) If $B_n \in \mathbf{B}$, then $\bigcup B_n \in \mathbf{B}$,
($\mathbf{B}$.3) If $B \in \mathbf{B}$, then $B^c = \Omega - B \in \mathbf{B}$.

An element of $\mathbf{B}$ is called an *event*.

Assume that associated with each B in $\mathbf{B}$ is a real number $P(B)$ satisfying the following properties:

(P.1) For any B in $\mathbf{B}$, $0 \le P(B) \le 1$,
(P.2) If $B_n \in \mathbf{B}$ are disjoint (i.e., $B_j \cap B_k = \emptyset$ for $j \neq k$), the equality $P(\bigcup B_n) = \sum P(B_n)$ holds.

Then, P is called a *probability measure* and $P(B)$ is said to be the *probability* of (the event) B.

The triple $(\Omega, \mathbf{B}, P)$ given above is called a *probability space*. The set Ω, which is an abstract set, is often written as $\Omega(\mathbf{B}, P)$ to emphasize the fact that $\mathbf{B}$ and P are attached. For an even B, its complement $B^c = \Omega - B$ is called the *complementary event* of B, according to the traditional terminology. If the events B_i, $i = 1, 2, \dots$, are such that $B_i \cap B_j = \varnothing$ $(i \neq j)$, then they are said to be mutually *exclusive.*

The independence of event is defined as follows. Two events A and B are called *independent* if

$$P(A \cap B) = P(A)P(B).$$

This notion is extended to a family of events, A_λ, $\lambda \in \Lambda$. If for any choice of finite number A_λ, say $A_{\lambda_1}, A_{\lambda_2}, \dots, A_{\lambda_n}$,

$$P\left(\bigcap A_{\lambda_k}\right) = \prod_k P(A_{\lambda_k}),$$

then the A_λ are called independent.

A family of σ-fields $\mathbf{B}_\lambda$, $\lambda \in \Lambda$, is called independent if for any choice of A_λ s from $\mathbf{B}_\lambda$, the A_λ, $\lambda \in \Lambda$, are independent.

A probability space $(\Omega, \mathbf{B}, P)$ is a measure space with, $P(\Omega) = 1$, as was defined before, and many notions in the theory of measure and integration are therefore diverted to those in probability theory, while probability theory often prefers its own traditional terminologies in order to maintain the intutitive meanings. A measurable function $X(\omega)$ on $(\Omega, \mathbf{B}, P)$ is by preference called a *random variable.* In case where X is vector-valued, we call X a *random vector.* If given random variables are indexed by a real variable, say $t \in T$, T an interval $(\subset R)$, or indexed by an integer, then they are denoted by $X(t, \omega)$ or by $X_t(\omega)$, and the family $X = \{X(t, \omega)\}$ of these random variables is called a *stochastic process.*

If a random variable $X(\omega)$ is integrable on Ω, then its *mean* (or *expectation*) is given by the integral $\int X(\omega)P(d\omega)$ and is denoted by $E(X)$. Further, if $|X(\omega)|^p$, $p \geq 1$, is integrable, then the *pth order moment* around the mean is defined by

$$\int_\Omega |X(\omega) - E(X)|^p \, P(d\omega) = E[|X - E(X)|^p].$$

In particular, the second order moment of X is called the *variance* of X and is usually denoted by $V(X)$. In case X is vector valued, its mean and moments are defined component wise, and we can introduce the *mean vector* and *covariance matrix* $V(X) = (V_{i,j})$ with entries

$$V_{i,j} = E\{(X_i - E(X_i))(X_j - E(X_j))\}.$$

Obviously V is symmetric and nonnegative definite.

Let $\{X_i(\omega);\ i = 1, 2, \dots\}$ be a system of random variables. The X_i are said to be (*mutually*) *independent,* if the $\mathbf{B}(X_i)$ are independent, where the symbol $\mathbf{B}(X)$ denotes the smallest σ-field with respect to which X is measurable. A similar definition is given even in the case where the X_i are vector valued.

PROPOSITION 1.1. *If random variables* $X_i(\omega)$, $i=1, 2, \ldots$, *on* $(\Omega, \mathbf{B}, P)$ *are independent, then the covariance matrix is of diagonal form.*

§2. Probability distributions of random variables (random vectors) and characteristic functions

First, we take a real-valued random variable $X(\omega)$, which is, by definition, a measurable mapping from $(\Omega, \mathbf{B}, P)$ to R, so that for any Borel subset Γ of R, the set $\{\omega; X(\omega) \in \Gamma\}$ (abbreviated $(X \in \Gamma)$) is a $\mathbf{B}$-measurable subset of Ω, that is, an event. Hence, $X(\omega)$ defines a probability measure $\Phi(\Gamma) = P(\{\omega; X(\omega) \in \Gamma\})$, (abbreviated $P(X \in \Gamma)$), $\Gamma \in \Sigma$, where Σ is the ordinary Borel field. A probability space (R, Σ, Φ) is thus introduced. This measure Φ on R is called the *probability distribution* of X. The function $F(x)$ on R defined by

$$F(x) = \Phi((-\infty, x]) \tag{1.1}$$

is called the *distribution function* of X. It is easy to see that F satisfies the following three conditions:

$(F.1)$ $F(x)$ is right continuous,
$(F.2)$ $F(x)$ is non-decreasing,
$(F.3)$ $\lim_{x\to\infty} F(x) = 1$, $\lim_{x\to-\infty} F(x) = 0$.

Conversely, a function $F(x)$ satisfying the conditions $(F.1)$–$(F.3)$ uniquely determines a probability measure Φ on R and we have a probability space (R, Σ, Φ).

If the measure Φ is absolutely continuous with respect to Lebesgue measure, then the Radon-Nikodym derivative $f(x) = F'(x)$ exists and is called the *density function* of the distribution Φ.

For the readers convenience we state the Radon-Nikodym Theorem, since it is often referred to in what follows.

THEOREM 1.1 (Radon-Nikodym). *Let two measures* μ *and* $\tilde{\mu}$ *be given on a measurable space* $(\Omega, \mathbf{B})$. *If* $\tilde{\mu}(A) = 0$ *implies* $\mu(A) = 0$ *for* $A \in \mathbf{B}$ *(namely,* μ *is absolutely continuous with respect to* $\tilde{\mu}$*), then there exists a nonnegative function* $f(\omega)$ *such that for any* $A \in \mathbf{B}$

$$\mu(A) = \int_A f(\omega)\, \tilde{\mu}(d\omega).$$

The function f *is uniquely determined on* Ω *up to a set of* μ*-measure zero.*

The function $f(\omega)$ thus obtained is the Radon-Nikodym derivative.

Let $\varphi(z)$ be the Fourier-Stieltjes transform of Φ:

$$\varphi(z) = \int_R \exp[izx]\, dF(x), \qquad z \in R. \tag{1.2}$$

Then $\varphi(z)$ is called the *characteristic function* of Φ (or of the distribution function of F). If, in particular, the measure Φ is the probability distribution of a random variable $X(\omega)$, then φ is also called the characteristic function of X and we have

$$\varphi(z) = E\{\exp[izX]\}, \qquad z \in R.$$

The following properties of $\varphi(z)$ can easily be proved.

(C.1) φ is positive definite: For any $z_1, z_2, \ldots, z_n \in R$ and for any $\alpha_1, \alpha_2, \ldots, \alpha_n \in \mathbf{C}$, we have $\sum \alpha_j \overline{\alpha}_k \varphi(z_j - z_k) \geq 0$.

(C.2) $\varphi(z)$ is uniformly continuous in z.

(C.3) $\varphi(0) = 0$.

Conversely, we can prove the following theorem.

THEOREM 1.2 (S. Bochner). *Let $\varphi(z)$ be a function of $z \in R$ satisfying the conditions* (C.1), (C.2) *and* (C.3). *Then there is a probability distribution Φ on the measurable space (R, Σ) such that*

$$\varphi(z) = \int_R \exp[izx]\,\Phi(dx).$$

More practically, the following theorem establishes the one-to-one correspondence between φ and Φ.

THEOREM 1.3 (P. Lévy's inversion formula). *Let $\varphi(z)$, $z \in R$, be the characteristic function of a distribution Φ. Then, we have*

$$\int_{-\infty}^{\infty} \chi_{\langle a, b\rangle}(x)\,\Phi(dx) = \lim_{c\to\infty} \frac{1}{2\pi} \int_{-c}^{c} \frac{\exp[-ibz] - \exp[-iaz]}{-iz} \varphi(z)\,dz,$$

where

$$\chi_{\langle a, b\rangle}(x) = \begin{cases} 1 & \textit{for } x \in (a, b), \\ 1/2 & \textit{for } x = a \textit{ or } b, \\ 0 & \textit{for } x \notin [a, b]. \end{cases}$$

Summing up, we can see the following bijective correspondences

$$\varphi \leftrightarrow F \leftrightarrow \Phi.$$

A sequence Φ_n, $n = 1, 2, \ldots$, of distributions on R is said to converge to a distribution Φ, if for any f in the space $\mathbf{C}_0 = \{f ; f$ is continuous and $\lim_{x\to-\infty} f(x) = \lim_{x\to\infty} f(x) = 0\}$ the integral $\int f(x)\Phi_n(dx)$ converges to $\int f(x)\Phi(dx)$. This concept of convergence of measures can be paraphrased in terms of characteristic functions as is shown in the following theorem.

THEOREM 1.4 (P. Lévy and V. Glivenko). (1) *If the distribution Φ_n converge to a distribution Φ, then the characteristic functions $\varphi_n(z)$ of Φ_n converge to the characteristic function of Φ uniformly on compact sets.*

(2) *If a sequence of the characteristic functions $\varphi_n(z)$ of distributions Φ_n converges to a function $\varphi(z)$ uniformly on some neighborhood of $z = 0$, then the limit $\varphi(z)$ is the characteristic function of some distribution Φ and Φ_n converges to Φ.*

What we have discussed so far in this section can easily be generalized to the case of multidimensional probability distributions and of real random vectors. We shall now briefly discuss the relation between distributions and characteristic functions with special emphasis on the difference from the one-dimensional case.

Let $\mathbf{X}(\omega)$ be an r-dimensional random vector $(1 < r < \infty)$, and set $\Phi(A) = P(\mathbf{X}(\omega) \in A)$, A being a Borel subset of R^r. Then we have a probability measure space (R^r, Σ, Φ), where Σ is the ordinary Borel field. Thus obtained Φ is the probability distribution of X and defines the distribution function F:

$$F(\mathbf{x}) = F(x_1, x_2, \dots, x_r) = \Phi\left(\prod_{k=1}^{r}(-\infty, x_k]\right), \qquad \mathbf{x} = (x_1, x_2, \dots, x_r),$$

which satisfies properties $(F.1)$ and $(F.2)$ for each x_k, $(k = 1, 2, \dots, r)$. While $(F.3)$ should be replaced by

$$\lim_{x_1, x_2, \dots, x_r \to \infty} F(\mathbf{x}) = 1,$$
$$\lim_{x_k \to -\infty} F(x_1, \dots, x_k, \dots, x_r) = 0 \quad \text{for fixed } x_j\ (j \neq k). \tag{$F.3'$}$$

In addition, we have

$$\begin{aligned} F(x_1 + h_1, x_2 + h_2, \dots, x_r + h_r) &- F(x_1, x_2 + h_2, \dots, x_r + h_r) - \cdots \\ - F(x_1 + h_1, x_2 + h_2, \dots, x_r) &+ F(x_1, x_2, x_3 + h_3, \dots, x_r + h_r) + \cdots \\ + (-1)^r F(x_1, x_2, \dots, x_r) &\geq 0, \qquad h_1, h_2, \dots, h_r \geq 0. \end{aligned} \tag{$F.3''$}$$

As for a characteristic function, we take $\mathbf{z} = (z_1, z_2, \dots, z_r)$ in R^r and define

$$\varphi(\mathbf{z}) = \int_{R^r} \exp[i(\mathbf{z}, \mathbf{x})]\, \Phi(dx), \qquad (\mathbf{z}, \mathbf{x}) = \sum_{j=1}^{r} x_j z_j.$$

We can now show some results similar to Theorem 1.1–1.3; in particular, we have a generalization of Theorem 1.3 stated as follows:

$$\begin{aligned} &\int_{R^r} \prod_{k=1}^{r} \chi_{(a_k, b_k)}(x_k)\, \Phi(d\mathbf{x}) \\ &\quad = \lim_{c \to \infty} \left(\frac{1}{2\pi}\right)^r \int_{[-c, c]^r} \prod_{k=1}^{r} \frac{\exp[-ib_k z_k] - \exp[-ia_k z_k]}{-iz_k} \\ &\qquad\qquad \cdot \varphi(z_1, \dots, z_r)\, dz_1 \cdots dz_r. \end{aligned}$$

The next theorem gives a condition for random vectors to be independent in terms of characteristic functions.

THEOREM 1.5. *Let $X_k(\omega)$ $(k = 1, 2, \dots, n)$ be r_k-dimensional random vectors. A necessary and sufficient condition for the X_k to be independent is that*

$$E\left\{\prod_{k=1}^{n} \exp[i(X_k, \mathbf{z}_k)]\right\} = \prod_{k=1}^{n} E\{\exp[i(X_k, \mathbf{z}_k)]\}$$

for any $\mathbf{z}_k \in R^{r_k}$.

In order to go further into the infinite dimensional analogue, additional background is necessary, as one can easily understand from the fact that difficulties arise from integration theory in infinite dimensional space. Noting this fact we shall discuss, in §6, probability distributions of infinite dimensional random vectors by restricting our attention to the particular case of a stochastic process, which may be viewed as a system of infinitely many random variables indexed by a time variable; it is infinite dimensional. Before coming to this topic, we present some limit theorems.

§3. Convergence of sequences of random variables

Various kinds of convergence of a sequence of random variables are introduced. Each of them is used depending on the purpose. The relation among them is, by itself, also interesting. Four different kinds of convergence are listed below.

(A) Let Φ_n be the distribution of a real-valued random variable $X_n(\omega)$, $n = 1, 2, \ldots$. If the sequence Φ_n converges to Φ with Φ the distribution of a random variable $X(\omega)$, then $X_n(\omega)$ is said to *converge to* $X(\omega)$ *in law*.

(B) If a sequence $X_n(\omega)$, $n = 1, 2, \ldots$, converges to $X(\omega)$ for P-almost all ω, then $X_n(\omega)$ is said to converge to $X(\omega)$ *almost surely* (or with probability one), and we write

$$\lim_{n\to\infty} X_n(\omega) = X(\omega) \quad \text{a.s. } P.$$

(C) If for any $\varepsilon > 0$

$$\lim_{n\to\infty} P(|X_n(\omega) - X(\omega)| > \varepsilon) = 0,$$

then $X_n(\omega)$ is said to converge to $X(\omega)$ *in probability*.

(D) Let $p \geq 1$, and assume that both $X_n(\omega)$ and $X(\omega)$ have pth order moment and that

$$\lim_{n\to\infty} E(|X_n - X|^p) = 0.$$

Then $X_n(\omega)$ is said to *converge to* $X(\omega)$ *in the* p*th mean* (or in *the* L^p*-sense*). If, in particular, $p = 2$, we simply say $X_n(\omega)$ is *convergent in the mean* and we write $\text{l.i.m.}_{n\to\infty} X_n = X$ (l.i.m. means "limit in the mean"). This concept is important in the study of Gaussian processes as we will see in Chapter 2.

The following relations among the above four different types of convergence can be proved.

(1) If $X_n(\omega)$ converges to $X(\omega)$ almost surely, then it also converges in probability.
(2) Convergence in probability implies convergence in law.
(3) Convergence in probability follows from L^p-convergence $(p \geq 1)$.
(4) If $p \geq q$, L^p-convergence implies L^q-convergence.

(5) If $X_n(\omega)$ converges in probability, then there exists a subsequence that converges almost surely.

§4. Limit theorems for sums of independent random variables

Throughout this section we assume that $X_n(\omega)$, $n = 1, 2, \dots$, is a sequence of independent random variables on $(\Omega, \mathbf{B}, P)$. The partial sum is denoted by $S_n(\omega)$, where

$$S_n(\omega) = \sum_{k=1}^{n} X_k(\omega).$$

Proposition 1.2. (1°) (Tchebychev's inequality). *If $X(\omega)$ has finite variance $V(X)$, then we have*

$$P(|X - E(X)| \geq t) \leq V(X)/t^2.$$

(2°) (Kolmogorov's inequality). *Assume that each random variable $X_n(\omega)$ has finite variance. Then, we have*

$$P\left(\max_{1\leq k\leq n} |S_k - E(S_k)| \geq t\right) \leq V(S_n)/t^2.$$

The following theorem is proved by Tchebychev's inequality.

Theorem 1.6 (The weak law of large numbers). *Assume that each X_k has finite variance $V(X_k)$ and that $V(X_k) \leq V$. Set $E(X_k) = m_k$. Then we have for any $\varepsilon > 0$*

$$\lim_{n\to\infty} P\left(\left|\frac{1}{n}\left(S_n - \sum_{k=1}^{n} m_k\right)\right| > \varepsilon\right) = 0.$$

By using Kolmogorov's inequality we can prove

Theorem 1.7 (The strong law of large numbers). *With the same assumptions as in the theorem above*

$$\lim_{n\to\infty} \frac{1}{n}\left\{S_n - \sum_{k=1}^{n} m_k\right\} = 0 \quad a.e.\ P.$$

Remark. The factor $1/n$ in the above equation may be replaced by $n^{-\alpha}$ with $\alpha = 1/2 + \varepsilon$, $\varepsilon > 0$. Namely, we have

$$\lim \frac{1}{n^\alpha}\left(S_n - \sum_{k=1}^{n} m_k\right) = 0 \quad \text{a.e. } P. \tag{1.3}$$

More precisely, the law of the iterated logarithm can be proved (see, e.g. A. N. Shirjaev 1984).

THEOREM 1.8 (Central Limit Theorem). *Assume that each* X_k *has finite variance* $V(X_k)$, *and set* $\sigma_n = \{\sum_{k=1}^n V(X_k)\}^{1/2}$ *and* $E(X_k) = m_k$. *Let* Φ_n *be the distribution of* X_n. *If the condition*

$$\sigma_n^{-2} \sum_{k=1}^n \int_{-\varepsilon\sigma_n}^{\varepsilon\sigma_n} x^2 \Phi_n(dx) \to 1 \quad \text{as } n \to \infty, \tag{$*$}$$

for any $\varepsilon > 0$, *then the distribution of*

$$\left\{S_n - \sum_{k=1}^n m_k\right\} / \sigma_n$$

converges to the standard Gaussian distribution.

REMARK. The condition $(*)$ is known as the *Lindeberg condition*, which seems to be a favorable condition sufficient for many applications.

There are many variations of the central limit theorem; simpler or more general. More general sufficient (and necessary) conditions can be given under somewhat different settings as is shown below (also, see A. N. Shirjaev 1984).

THEOREM 1.8′. *Suppose that for any* $n \geq 1$ *we have a sequence*

$$X_{n,1}, X_{n,2}, \ldots, X_{n,n}$$

of independent random variables with $E(X_{n,i}) = 0$ *and* $V(X_{n,i}) = \sigma_{n,i}^2 > 0$ *for* $i = 1, 2, \ldots, n$, *and* $\sum_{i=1}^n \sigma_{n,i}^2 = 1$.

Set $S_n = X_{n,1} + X_{n,2} + \cdots + X_{n,n}$, *and set* $F_{n,i}(x) = P(X_{n,i} \leq x)$. *Then, a necessary and sufficient condition for the distribution* S_n *to converge to the standard Gaussian distribution* $N(0, 1)$ *is that*

$$\sum_{i=1}^n \int_{|x|>\varepsilon} |x|\,|F_{n,i}(x) - \Phi(x/\sigma_{n,i})|\,dx \to 0 \quad \text{as } n \to \infty$$

for any $\varepsilon > 0$.

§5. Conditional expectations and martingales

Conditional expectations are defined by using the Radon-Nikodym theorem for the basic probability space. The following theorem shows how to define a conditional expectation. Let a probability space $(\Omega, \mathbf{B}, P)$ be given.

THEOREM 1.9. *For a* P*-integrable function* $f(\omega)$ *a (signed) measure* μ *is defined by*

$$\mu(A) = \int_A f(\omega)\, P(d\omega), \qquad A \in \mathbf{B}.$$

Let $\widetilde{\mathbf{B}}$ *be a sub-*σ*-field of* $\mathbf{B}$ *and let* $\tilde{\mu}$ *be the restriction of* μ *to* $\widetilde{\mathbf{B}}$. *Then, there is a unique* $\widetilde{\mathbf{B}}$*-measurable function* $g(\omega)$ *such that*

$$\tilde{\mu}(B) = \int_B g(\omega)\, P(d\omega), \qquad B \in \widetilde{\mathbf{B}}. \tag{1.4}$$

The proof is based on the fact that $\tilde{\mu}$ is absolutely continuous with respect to P restricted to $\widetilde{\mathbf{B}}$.

DEFINITION 1.1. The $\widetilde{\mathbf{B}}$-measurable function $g(\omega)$ determined by formula (1.4) is called the *conditional expectation* of $f(\omega)$ relative to the σ-field $\widetilde{\mathbf{B}}$ and is denoted by $E(f|\widetilde{\mathbf{B}})$. If, in particular, $f(\omega)$ is the indicator function $\chi_A(\omega)$ of a set $A \in \mathbf{B}$ (i.e., $\chi_A(\omega) = 1$ for $\omega \in A$, $= 0$ for $\omega \in A^c$), then the conditional expectation $E(\chi_A|\widetilde{\mathbf{B}})$ is called the *conditional probability* of A relative to the σ-field $\widetilde{\mathbf{B}}$ and is denoted by $P(A|\widetilde{\mathbf{B}})$.

By definition, we have

$$E(E(f|\widetilde{\mathbf{B}})) = E(f) \quad \text{and} \quad E(P(A|\widetilde{\mathbf{B}})) = P(A).$$

If $\widetilde{\mathbf{B}}$ is the σ-field generated by a system of random variables $\{X_i,\ i = 1, 2, \dots\}$, then $E(f|\widetilde{\mathbf{B}})$ is often denoted by $E(f|X_i,\ i = 1, 2, \dots)$. An obvious relation $E(f|\mathbf{B}_0) = E(f)$ hold for the trivial σ-field $\mathbf{B}_0 = \{\varnothing, \Omega\}$. The next assertions follow easily from the definition of conditional expectation.

PROPOSITION 1.3. (1) *For any constants* a, b *we have*

$$E(af_1 + bf_2|\widetilde{\mathbf{B}}) = aE(f_1|\widetilde{\mathbf{B}}) + bE(f_2|\widetilde{\mathbf{B}}),$$

(2) *If* $f_n(\omega)$ (≥ 0), $n = 1, 2, \dots$, *is an increasing sequence and converges to* $f(\omega)$, *then* $E(f_n|\widetilde{\mathbf{B}})$ *converges to* $E(f|\widetilde{\mathbf{B}})$.

(3) *For two* σ*-fields* $\mathbf{B}_1$ *and* $\mathbf{B}_2$ *such that* $\mathbf{B}_1 \supset \mathbf{B}_2$, *we have*

$$E[E(f|\mathbf{B}_1)|\mathbf{B}_2] = E(f|\mathbf{B}_2). \tag{1.5}$$

(4) *If two* σ*-fields* $\mathbf{B}_1$ *and* $\mathbf{B}_2$ *are independent, then*

$$E(f_1|\mathbf{B}_2) = E(f_1)$$

for any integrable $\mathbf{B}_1$*-measurable function* f_1.

(5) *If* $f_2(\omega)$ *is* $\mathbf{B}$*-measurable, we have*

$$E(f_1 f_2|\mathbf{B}) = f_2 E(f_1|\mathbf{B}).$$

REMARK. The equalities in Proposition 1.3 should hold with probability one. If one wishes to state this fact explicitly, the symbol a.e. P, should follow right after each equality.

We use the notion of conditional expectation to introduce a particular stochastic process, called a martingale.

DEFINITION 1.2. Let a system of random variables $M = \{M_n(\omega);\ n = 1, 2, \dots\}$ and an increasing sequence of sub-σ-fields $\{\mathbf{B}_n;\ n = 1, 2, \dots\}$ of $\mathbf{B}$ be given. Assume that the following three conditions M.1–3) are satisfied:

(M.1) M_n is $\mathbf{B}_n$-measurable, $n = 1, 2, \dots$,
(M.2) $E(|M_n|) < \infty$, $n = 1, 2, \dots$,
(M.3) $E(M_n|\mathbf{B}_m) = M_m$ for any $n > m$ a.e. P.

Then, $M = \{M_n, \mathbf{B}_n;\ n = 1, 2, \dots\}$ is called a (discrete parameter) *martingale.*

If the equality in $(M.3)$ is replaced with the inequality $\geq$ (or $\leq$), then M is called *sub* (or *super*) *martingale.*

REMARK 1. A martingale M is surely an ordinary stochastic process which may be denoted simply by M_n, however a martingale it is better to denote it by $\{M_n, \mathbf{B}_n;\ n = 1, 2, \dots\}$ as above, since the family of σ-fields plays an essential role and is worth writing out explicitly.

REMARK 2. In the above definition the time parameter set is taken to be the set N of all natural numbers, however any linearly ordered set can also be taken when we define a martingale. One of the sets: $N \cup \{\infty\}$, the intervals $(0, \infty)$, $(-\infty, \infty)$, or $[0, 1]$ will often be taken in what follows.

Three examples of a martingale are now in order.

EXAMPLE 1. Let $f(\omega)$ be a random variable such that $E(|f(\omega)|) < \infty$. Given an increasing sequence of sub-σ-fields $\{\mathbf{B}_n;\ n = 1, 2, \dots\}$ of $\mathbf{B}$. Then, by setting $M_n(\omega) = E(f|\mathbf{B}_n)$, we obtain a martingale

$$M = \{M_n, \mathbf{B}_n;\ n = 1, 2, \dots\}.$$

A martingale which is obtained from the conditional expectations of a random variable, as in Example 1, is said to be *regular.*

EXAMPLE 2. Let $\{X_n(\omega);\ n, 2, \dots\}$ be a sequence of independent random variables with mean zero. Set

$$\mathbf{B}_n = \mathbf{B}_n(X) = \{\text{the sub-}\sigma\text{-field generated by } X_1, X_2, \dots, X_n\}$$

and set

$$S_n(\omega) = \sum_{i=1}^{n} X_n(\omega), \qquad n = 1, 2, \dots.$$

Then $\mathbf{S} = \{S_n, \mathbf{B}_n;\ n = 1, 2, \dots\}$ is a martingale.

EXAMPLE 3. Take a martingale $M = \{M_n, \mathbf{B}_n;\ n = 1, 2, \dots\}$ and a convex function $f(x)$, $x \in R$. Then $f(M) = \{f(M_n), \mathbf{B}_n;\ n = 1, 2, \dots\}$ is a submartingale. This property can be proved by using the Jensen inequality.

DEFINITION 1.3. Let $\tau(\omega)$ be an N-valued random variable and let $\{\mathbf{B}_n;\ n = 1, 2, \dots\}$ be an increasing sequence of sub-σ-fields of $\mathbf{B}$. If for any $n \in N$ the set $\{\omega;\ \tau(\omega) \leq n\}$ is an element of $\mathbf{B}_n$, then τ is called a *stopping time* relative to the $\{\mathbf{B}_n\}$.

Set $\mathbf{B}_{n\wedge\sigma} = \{A \in \mathbf{B};\ A \cap \{n \wedge \sigma \leq m\} \in \mathbf{B}_m,\ m \in N\}$ with a stopping time σ and with the notation $a \wedge b = \min\{a, b\}$. Then $\mathbf{B}_{n\wedge\sigma}$ is again a σ-field. The following assertion is straightforward from the definition.

PROPOSITION 1.4. *Let $\tau(\omega)$ and $\sigma(\omega)$ be stopping times such that $\tau(\omega) \geq \sigma(\omega)$ holds a.e. P. If $M = \{M_n, \mathbf{B}_n;\ n = 1, 2, \dots\}$ is a martingale, then*

$$M_{n\wedge\sigma} = E[M_{n\wedge\tau}|\mathbf{B}_{n\wedge\sigma}] \quad \text{a.e. } P. \tag{1.6}$$

We now make a few remarks.

REMARK. (i) If σ is a stopping time, so is $\sigma \wedge n$.

(ii) For a stopping time σ, the σ-field $\mathbf{B}_\sigma$, as it were $\mathbf{B}_{\infty\wedge\sigma}$, itself is defined by $\{A \in \mathbf{B};\ A \cap \{\sigma \le m\} \in \mathbf{B}_m\}$.

(iii) If M in Proposition 1.4 is a sub (or super) martingale, then the equality in (1.6) should be replaced by the inequality $\ge$ (or $\le$).

A martingale can be thought of as a generalization of a sum of independent random variables, and therefore the following theorem seems to be a reasonable generalization to a martingale of Kolmogorov's inequality (see Proposition 1.2.).

THEOREM 1.10 (Doob's inequality). *Let M be a martingale (or submartingale). Then for any N and for any $\lambda > 0$, we have*

$$P\left(\max_{n\le N} M_n \ge \lambda\right) \le \frac{1}{\lambda}\int_{\{\max_{n\le N} M_n \ge \lambda\}} M_N(\omega)\, P(d\omega) \le \frac{1}{\lambda} E(M_N^+),$$

where $M_N^+ = \max\{M_N, 0\}$, and we have

$$P\left(\min_{n\le N} M_n \le -\lambda\right) \le \frac{1}{\lambda} - E(M_1) + \left\{\int_{\{\min_{n\le N} M_n > -\lambda\}} M_N(\omega)\, P(d\omega)\right\}.$$

Asymptotic behaviors of a martingale as $N \to \infty$ are particularly important. The next result will often be used in the discussion that follows.

THEOREM 1.11. *Let $M = \{M_n, \mathbf{B}_n;\ n = 1, 2, \ldots\}$ be a martingale. Then we have*

$$E(|M_1|) \le E(|M_2|) \le \cdots \le E(|M_n|) \le \cdots.$$

If $\lim_{n\to\infty} E(|M_n|) = K < \infty$, then

$$\lim_{n\to\infty} M_n(\omega) = M_\infty(\omega)$$

exists with probability one.

COROLLARY. *Let M be a regular martingale (see Example 1). Then the limit*

$$\lim_{n\to\infty} M_n(\omega) = M_\infty(\omega) \tag{1.7}$$

always exists with probability one, and the new system

$$\overline{M} = \{M_n, \mathbf{B}_n;\ n = 1, 2, \ldots, \infty\}$$

is again a martingale, where $\mathbf{B}_\infty = \sigma\{\mathbf{B}_n;\ n = 1, 2, \ldots, \infty\}$ is the σ-field generated by all the $\mathbf{B}_n$, $n \ge 1$.

Conversely, if for a martingale M the limit (1.7) exists and if $\overline{M}$ is a martingale, then M is regular.

These results are useful when we discuss, in Chapter 6, the equivalence problem for Gaussian processes. Also note that if M is regular, then from the expression $M_n = E(f\,|\mathbf{B}_n)$ we see that $M_\infty = E(f\,|\mathbf{B}_\infty)$.

§6. Measures on function spaces (measures induced from stochastic processes)

As we explained in §1, a stochastic process $\{X(t, \omega);\ t \in T\}$ is a system of random variables indexed by a linearly ordered set T. Once an ω in Ω is fixed, a sample function is given; we denote it by $X(t, \omega)$. It is a function of t. Thus, we are given a mapping from Ω to R^T. If one wishes to make a stochastic process concretized, then first Ω is taken to be R^T, then a probability measure μ should be introduced on R^T with a suitable σ-field so that any finite-dimensional restriction of μ is in agreement with a certain marginal distribution of the probability distribution of $\{X(t, \omega);\ t \in T\}$.

The first step in our approach is, therefore, to establish a σ-field of subsets of R^T which is the set of all real-valued functions on T. A subset A of R^T is said be a *cylinder set* if there exist a finite number, say n $(n < \infty)$, of t_j, $j = 1, 2, \dots, n$, in T and a Borel subset $B \subset R^n$ such that A is expressed in the form

$$A = \{x \in R^T;\ (x(t_1), x(t_2), \cdots, x(t_n)) \in B\}. \tag{1.8}$$

An obvious remark to be mentioned is that the representation (1.8) is not unique for a given A.

The smallest σ-field involving all the cylinder sets will be denoted by $\mathbf{B}^T$, and this σ-field is exactly what we propose to take. Here is a recipe to construct it: for fixed $t_1, t_2, \dots, t_n$, a σ-field $\mathbf{B}^{(t_1, t_2, \dots, t_n)}$ is generated by the sets of the form (1.8), where the n-tuple $(t_1, t_2, \dots, t_n)$ is fixed. Obviously, if $(t_1, t_2, \dots, t_n)$ is a subset of another m-tuple $(s_1, s_2, \dots, s_m)$, then $\mathbf{B}^{(t_1, t_2, \dots, t_n)} \subset \mathbf{B}^{(s_1, s_2, \dots, s_m)}$. The lattice sum of such σ-fields defines $\mathbf{B}^T$.

The following theorem is well known (see, e.g. A. N. Shirjaev, 1984).

THEOREM 1.12 (A. N. Kolmogorov). *If a real-valued set function $\mu_0(A)$ is defined for any cylinder set A, and if for any fixed t_j, $1 \le j \le n$, μ_0 is a probability measure on the σ-field $\mathbf{B}^{(t_1, \dots, t_n)}$, then there exists a probability measure μ on $\mathbf{B}^T$ such that μ is a unique extension of μ_0.*

As a corollary, we conclude that for a stochastic process $X = \{X(t); t \in T\}$ on a probability space $(\Omega, \mathbf{B}, P)$ a set function μ_0 is introduced by

$$\mu_0(A) = P(\omega;\ (X(t_1, \omega), X(t_2, \omega), \dots, X(t_n, \omega)) \in B)$$

for a cylinder set A given by (1.8) and that a probability measure μ is given on $(R^T, \mathbf{B}^T)$ as the unique extension of μ_0.

The measure μ thus obtained is said to be the *distribution* on R^T introduced by a stochastic process X, or simply the *distribution* of X.

Conversely, suppose that a system $M = \{\mu^{(t_1, \dots, t_n)};\ t_1, \dots, t_n \in T,\ n \in N\}$ is given, and assume the *consistency condition* for M in the following sense: Consider the case where $\{t_1, \ldots, t_n\} \subset \{s_1, \ldots, s_m\}$, $n < m$, where without loss of generality, we can assume that $t_1 = s_1, \ldots, t_n = s_n$. Then

$$\mu^{(t_1, \ldots, t_n)}(B) = \mu^{(t_1, \ldots, t_n, \ldots, s_m)}(B \times R^{m-n}), \qquad B \subset R^n. \tag{1.9}$$

With this assumption we can define μ_0 in such a way that for a cylinder set A of the form (1.8) we set $\mu_0(A) = \mu^{(t_1, \ldots, t_n)}(B)$. In fact, μ_0 is well defined on the family of cylinder sets. Now apply Theorem 1.12 to obtain a probability measure μ on $(R^T, \mathbf{B}^T)$ which is the extension of μ_0.

We are now ready to prove the existence theorem of a stochastic process when a consistent family of finite dimensional distributions is given.

THEOREM 1.13. *If a system M of finite dimensional probability distributions satisfies the consistency condition* (1.9), *then there is a stochastic process $X = \{X(t);\ t \in T\}$, the distribution of which is the measure μ on $(R^T, \mathbf{B}^T)$ formed above.*

Indeed, we may set $\Omega = R^T$, $\mathbf{B} = \mathbf{B}^T$, $P = \mu$, and $X(t, \omega) = x(t)$ with $x = \omega \in \Omega$ to prove the theorem.

It is an interesting question in probability to ask to $X(t, \omega)$, obtained in the above manner, is continuous in t for any ω in Ω. Unfortunately, the set $C = C(T)$ is not a $\mathbf{B}^T$-measurable subset of R^T. However, if the outer measure, derived from μ, of C is 1, then we may define a measure m on a σ-field $\mathscr{C}$ generated by the cylinder subsets of C in such a way that

$$m(\Gamma) = \mu(G) \quad \text{with } C \cap G = \Gamma,\ G \in \mathbf{B}^T.$$

In this case, a sample function of the stochastic process can be considered to be a member of C for almost all ω. The next theorem gives a criterion for a sample function of a stochastic process to be continuous in t.

THEOREM 1.14 (A. N. Kolmogorov and Y. V. Prokhorov). *If a stochastic process $X = \{X(t);\ t \in T\}$, given by Theorem* 1.13, *satisfies the condition*

$$E(|X(t) - X(s)|^a) \le c|t - s|^{1+b} \tag{1.10}$$

for some positive constants a, b, and c, then the outer measure relative to μ of the set C is equal to one, where μ is the distribution of X.

CHAPTER II

Systems of Gaussian Random Variables

In this chapter we deal with various systems of random functions involving infinitely many real or complex Gaussian random variables. One of the most significant properties of such a system is, intuitively speaking, that the basic analysis of the system can successfully be carried out as we shall see in §2, by applying linear operations because of its linear structure. The complexification of systems of real Gaussian random variables will be discussed in §3, and some particularly important roles played by complex Gaussian systems will be seen in the next chapter, where the spectral analysis for stationary Gaussian processes is discussed.

Some introductory interpretations of the canonical representation theory for Gaussian processes will be given in §§4–5 in a very naive manner. In particular, the technique of forming the *innovation* from a given Gaussian process is explained in §4, in the case of discrete parameter. In the continuous parameter case, the technique is quite complicated. The ideas follow along the same lines so that we will, in this chapter, present only simple, but typical, examples together with some background. Details of the theory will be discussed in Chapter III.

§1. Definitions and preliminaries

A real-valued random variable $X(\omega)$ is called *Gaussian* if its probability distribution is absolutely continuous and the density function is given by

$$\frac{1}{\sqrt{2\pi}\sigma}\exp\left[-\frac{(x-m)^2}{2\sigma^2}\right], \qquad \sigma > 0. \tag{2.1}$$

The constants m and σ^2 in the above expression are the *mean* and the *variance*, respectively. This distribution is denoted by $N(m, \sigma^2)$. If $m = 0$ and $\sigma = 1$, then X is said to be a *standard Gaussian* distribution.

An n-dimensional Gaussian distribution is defined by the density function

$$p(x) = \frac{1}{(2\pi)^{n/2}\sqrt{|V|}}\exp\left\{-\frac{1}{2}(V^{-1}(x-m), (x-m))\right\}, \qquad x \in R^n, \tag{2.2}$$

where m $(\in R^n)$ is the mean vector and where V is a positive definite

$n \times n$ matrix called the covariance matrix. If an n-dimensional Gaussian distribution with density function given by (2.2), then X is called an n-dimensional *Gaussian random vector*, where $m = E(X)$ is the mean vector and $V = (V_{i,j})$, $V_{i,j}$ being the covariance of X_i and X_j, is the covariance matrix of X.

We now come to a general definition.

DEFINITION 2.1. A system of random variables $X = \{X_\lambda ; \lambda \in \Lambda\}$ is called *Gaussian*, if any finite linear combination $\Sigma a_k X_{\lambda_k}$, $a_k \in R$, is a Gaussian random variable. If, in particular, X is a stochastic process, we call it a *Gaussian process*.

The mean vector $m = (m_\lambda ; \lambda \in \Lambda)$ and the covariance matrix $V = (V_{\lambda,\mu} ; \lambda, \mu \in \Lambda)$ are also defined in the case of a Gaussian system. In particular, for the case of a stochastic process, they are called the mean function and the covariance function, respectively. The covariance matrix V is positive definite in the sense that for any finite number of λ, say $\lambda_1, \lambda_2, \ldots, \lambda_n$, the $n \times n$ matrix $V = (V_{\lambda_i,\lambda_j})$ is positive definite. Note that such a property is still true for the covariance matrix of a general (not necessarily Gaussian) system of random variables with finite variance.

With the assumption that the system in question is Gaussian, we can prove the following remarkable theorem.

THEOREM 2.1. *Let $m = (m_\lambda ; \lambda \in \Lambda)$ be any vector and let $V = (V_{\lambda,\mu})$ be any nonsingular, symmetric positive definite matrix. Then, there exists a Gaussian system $X = \{X_\lambda ; \lambda \in \Lambda\}$, the mean vector and the covariance matrix of which are in agreement with the given m and V, respectively. In addition, the distribution of such an X is unique.*

REMARK. The last assertion means that if $X = \{X_\lambda ; \lambda \in \Lambda\}$ and $X' = \{X'_\lambda ; \lambda \in \Lambda\}$ are both Gaussian systems with same mean m and the same covariance matrix V, then for any finite number of λ, say $\lambda_1, \lambda_2, \ldots, \lambda_n$, the distribution of $(X_{\lambda_1}, X_{\lambda_2}, \ldots, X_{\lambda_n})$ and that of $(X'_{\lambda_1}, X'_{\lambda_2}, \ldots, X'_{\lambda_n})$ are the same.

OUTLINE OF THE PROOF OF THEOREM 2.1. First we form a required Gaussian system. Take Ω to be the function space R^Λ and let $\mathbf{B}$ be the σ-field generated by all the cylinder subsets of Ω to obtain a measurable space $(\Omega, \mathbf{B})$. For a cylinder subset $C = \{\omega = (\omega_\lambda); (\omega_{\lambda_1}, \omega_{\lambda_2}, \ldots, \omega_{\lambda_n}) \in B\}$ with B a Borel subset of R^n, we define

$$P_{\lambda_1, \lambda_2, \ldots, \lambda_n}(C) = \int_B p(x)\, dx, \tag{2.3}$$

where $p(x)$ is a density function like the one given by (2.2). Thus we have a probability measure on the σ-field involving cylinder subsets for any choices of finite number of λ. Then it is proved that the system of the measures obtained in this manner satisfies the consistency condition. We can therefore appeal to the Kolmogorov extension theorem (see Theorem 1.12.) to introduce a unique (completely additive) probability measure P on the measurable space $(\Omega, \mathbf{B})$. Finally, set $\omega_\lambda = X_\lambda(\omega)$. Then, we can see that $\{X_\lambda; \lambda \in \Lambda\}$ is a Gaussian system defined on the probability space $(\Omega, \mathbf{B}, P)$ with the mean vector m and the covariance matrix V given in advance.

The second half of the theorem can be proved by noting that any finite dimensional Gaussian distribution is uniquely determined by a mean vector and a covariance matrix. □

There is a short remark regarding the above theorem. Namely, we should note the case where the covariance matrix is singular for some choice of finite number of λ. Even in such a case we are still given a Gaussian distribution which is degenerated to some lower dimensional space.

The following theorem is often referred to, but we shall omit the proof, since it is readily found in ordinary text books.

THEOREM 2.2. (i) *Let X be a Gaussian system. Any subsystem of X is again Gaussian. If we add to X the limit of members in X in the sense of almost sure convergence or limit in the mean, or else limit in probability, then the new system is still a Gaussian system. The linear span of X is also Gaussian; the same for its closure.*

(ii) *Let $X = \{X_1, X_2, \dots, X_n, \dots\}$ be a Gaussian system. Assume that the X_n converge to X. Then, the sequence also converges in the mean.*

For a Gaussian system X, the closure of the linear span of X, which is Gaussian by the above theorem is a closed subspace of the Hilbert space $L^2(\Omega, P)$ and is denoted by $H(X)$. This space will often be used in the following sections.

§2. Some important properties of a Gaussian system

We are going to list some important characteristic properties of a Gaussian distribution and a Gaussian system (see, e.g., Hida 1970). The first four properties are concerned with one-dimensional Gaussian distributions.

(1) First, we introduce a general notion. Let $\psi(t)$ be the characteristic function of a one-dimensional distribution, and assume that $\log \psi(t)$ is expressed in the form

$$\log \psi(t) = \sum_{k=1}^{n} \frac{(it)^k}{k!} \gamma_k + o(t^n).$$

Then, the coefficient γ_k is called the kth *degree semi-invariant* of the distribution, and we say that it has semi-invariants up to the nth degree.

The characteristic function $\varphi(t)$ of a Gaussian distribution $N(m, \sigma^2)$ is given by

$$\varphi(t) = \exp\left[imt - \frac{\sigma^2}{2}t^2\right]. \tag{2.4}$$

Hence, we see that there exist semi-invariants of all degrees and that

$$\gamma_k = 0 \quad \text{for } k \geq 3. \tag{2.5}$$

Conversely, if a distribution has semi-invariants of all degrees and they satisfy the condition (2.5), then the distribution is either a Gaussian distribution or a δ-distribution.

Incidentally, for the Gaussian distribution $N(m, \sigma^2)$ we have

$$\gamma_1 = m, \qquad \gamma_2 = \sigma^2,$$

and the higher order moments m_k around the mean are

$$m_{2k+1} = 0, \qquad m_{2k} = (2k-1)!!\sigma^{2k}.$$

(2) Next we give an information theoretic characterization of the Gaussian distribution. Let $p(x)$ be the density function of a distribution on R. Then the amount of information contained in the distribution is measured by

$$H(p) = -\int_{-\infty}^{\infty} p(x)\log p(x)\,dx.$$

(Tacitly, $0\log 0$ is understood to be 0 in the above integral.) The $H(p)$ is obviously invariant under translation of the distribution. Among the distributions p with fixed variance σ^2, the maximum value of $H(p)$ is attained by the Gaussian distribution (m being arbitrary). This fact is proved by using the Jensen inequality, since $x\log x$ is a convex function on $[0, \infty)$. The exact value of $H(p)$ for Gaussian distribution with variance σ^2 is $\frac{1}{2} + \frac{1}{2}\log(2\pi\sigma^2)$.

(3) The Gaussian distribution is characterized as a stable distribution with exponent 2. Take $n+1$ independent identically distributed random variables $X_0, X_1, \ldots, X_n$, and form the sum $S_n = \sum_{i=1}^n X_i$. If S_n and $c_nX_0 + r_n$ are subject to the same distribution with a suitable choice of constants c_n and r_n, then the distribution is said to be *stable*. If $r_n = 0$ for every n, the distribution is called *strictly stable*.

It can be proved that $c_n = n^{1/\alpha}$ must hold for some α with $0 < \alpha \leq 2$. The power α is called the (characteristic) *exponent* of the stable distribution. The characteristic function φ_α of a symmetric stable distribution of exponent α can be written in the form

$$\varphi_\alpha(t) = \exp[-c|t|^\alpha], \qquad c > 0. \tag{2.6}$$

A Gaussian distribution can be considered as a stable distribution with exponent 2. In fact, if the density function is given by (2.1), then the sum S_n has mean nm and variance $n\sigma^2$, so that c_n and r_n in the above definition are taken to be $n^{1/2}$ and $(n-\sqrt{n})m$, respectively. This means that $\alpha = 2$. It is proved that a centered (i.e., $m = 0$) Gaussian distribution is a strictly stable distribution with exponent 2. Conversely, a strictly stable distribution with exponent 2 has to be a Gaussian distribution.

(4) The following method of characterizing Gaussian distributions is found in C. F. Gauss's paper (1809; see Gauss Werke Bd.VII, pp. 236–257), where he discusses celestial mechanics. The idea came from his theory of errors.

Let $p(x)$ be a density function of a distribution over a (finite or infinite) interval of R. In reality, $p(x)$ may be thought of as a density of error distribution for some measurement, so that we may assume that p is smooth enough and symmetric: $p(-x) = p(x)$. Suppose that we obtain a sample $\{x_1, x_2, \ldots, x_n\}$ of size n, and that we wish to have the best estimate of the mean m. Since $x_i - m$ is the error, we consider the likelihood in terms of p:

$$L(x, m) = \prod_{i=1}^{n} p(x_i - m).$$

Now assume that $p(x)$ is a density function such that the likelihood estimator of m is always given by the arithmetic mean of the samples. This seems to be a quite reasonable assumption for an error distribution. Hence, we see that for a given $x = (x_1, x_2, \ldots, x_n)$, $\log L(x, m)$, as a function of m, assumes its maximum at $\overline{x} = \frac{1}{n}\sum_{i=1}^{n} x_i$. It follows therefore, that

$$\sum_{i=1}^{n} q'(x_i - \overline{x}) = 0,$$

where $q(x) = \log p(x)$. If, in particular, $x_2 = x_3 = \cdots = x_n = a$, $x_1 = nx + a$, then we must have

$$q'((n-1)x) = (1-n)q'(-x) = (n-1)q'(x).$$

More generally, we can show that

$$q'(\lambda x) = \lambda q'(x), \qquad \lambda \text{ real}.$$

Hence, $q(x)$ has to be a polynomial in x of degree at most 2. Since $p(x)$ is a probability density function, the $p(x)$ is proved to be

$$p(x) = C \cdot \exp\left[-\frac{x^2}{V}\right], \qquad C, V > 0$$

on the interval in consideration, where C is a normalizing constant. In case the distribution extends over the entire R, $p(x)$ is of the form

$$p(x) = \frac{1}{\sqrt{2\pi}} \exp\left[-\frac{x^2}{2\sigma^2}\right].$$

(5) If $\{X, Y\}$ is a Gaussian system such that X and Y are not linearly dependent, then the conditional distributions are also Gaussian; for instance $P(X \le x|Y)$ is a Gaussian distribution. As for the conditional expectation, both $E(X|Y)$ and $E(Y|X)$ are linear in X and Y. These properties come

from the analytic property of the density function of the two-dimensional Gaussian distribution:

$$\frac{1}{2\pi\sigma\sigma'\sqrt{1-\rho^2}}\exp\left\{-\frac{1}{2(1-\rho^2)}\left\{\frac{(x-m)^2}{\sigma^2}-2\rho\frac{(x-m)(y-m')}{\sigma\sigma'}+\frac{(y-m')^2}{\sigma'^2}\right\}\right\}, \tag{2.7}$$

where

$$E(X)=m\,,\quad E(Y)=m'\,,\quad V(X)=\sigma^2\,,\quad V(Y)=\sigma'^2$$

and

$$E\{(X-m)(Y'-m')\}\equiv \operatorname{Cov}(X\,,\,Y)=\sigma\sigma'\rho\,.$$

The linear properties mentioned above can easily be extended to the case of Gaussian systems involving more than two variables. For example, given a Gaussian system $\{X\,,\,Y_1\,,\,Y_2\,,\,\dots\,,\,Y_n\}$, then the conditional expectation $E(X|Y_1\,,\,Y_2\,,\,\dots\,,\,Y_n)$ is, up to constant, a linear combination of the Y_i, $1\le i\le n$.

While if Y is Gaussian and $E(X|Y)$ is affine in Y and if the conditional distribution $P(X\le x|Y)$ is Gaussian almost surely, then $\{X\,,\,Y\}$ is a Gaussian system.

In general, if X and Y are independent and have finite variance, then their covariance (exists and) vanishes. If, in addition, $\{X\,,\,Y\}$ is a Gaussian system, the converse is true, that is, "covariance $=0$" implies "independence", as is easily seen from (2.7).

(6) Next we discuss the *reproducing property of Gaussian distribution.* If X and Y are independent and if each of them is Gaussian, then their sum $X+Y$ is again Gaussian in distribution as is easily seen. It is surprising that a partial converse is true. Namely, we have

THEOREM 2.3 (P. Lévy and H. Cramér). *If X and Y are independent and if their sum is Gaussian, then both X and Y are Gaussian.*

This assertion can be proved by using characteristic functions. In fact, we see by assumption that $X+Y$ is Gaussian, the characteristic functions of X and Y are both entire functions of exponential type of degree 2, which implies the conclusion. Detailed proof is in many references (see, e.g., T. Hida 1975).

(7) The next striking result is due to P. Lévy. Let $(X\,,\,Y)$ be a two-dimensional random vector. If there exist a random variable U, that is independent of X, and a random variable V, that is independent of Y, such that

$$Y=aX+U\,,\qquad X=bY+V\,, \tag{2.8}$$

then there are only the three possibilities listed below:

(i) $\{X\,,\,Y\}$ is a Gaussian system,

(ii) X and Y are independent,

(iii) there is a linear relation between X and Y; that is, there are constants α, β, γ such that $\alpha X + \beta Y = \gamma$.

For a proof and further discussions we refer to P. Lévy (1957).

REMARK. Looking at properties (5), (6), and (7), we can say that a Gaussian system enjoys a linear structure if an intuitive interpretation is permitted. It is for this reason that the Hilbert space theory provides a powerful tool.

§3. Complex Gaussian systems

We now come to systems of complex-valued Gaussian random variables. First of all, one may ask how to introduce a complex Gaussian system as a natural extension of a Gaussian system which enjoys the good properties shown in the last section. Certainly, it is not enough to simply assume that the real and imaginary parts are both Gaussian. Somewhat restricted, additional requirements for the system are necessary in order to define a reasonable class, the structure of which is similar to Gaussian systems discussed in the previous two sections.

We start with the case of a single random variable. Let $Z(\omega)$ be a complex-valued random variable on a probability space $(\mathbf{\Omega}, \mathbf{B}, P)$, where $Z(\omega)$ has the form

$$Z(\omega) = X(\omega) + iY(\omega) + m, \qquad m \text{ a complex number}, \tag{2.9}$$

with $X(\omega)$ and $Y(\omega)$ mutually independent, centered (i.e., mean $= 0$), Gaussian random variables with the same variance. Then, $Z(\omega)$ is called a *complex Gaussian random variable.* Obviously, $E(Z) = m$.

DEFINITION 2.2. Let $Z = \{Z_\lambda(\omega); \lambda \in \Lambda\}$ be a system of complex Gaussian random variables. If a linear combination (with complex coefficients) of any finite number of members in Z is always complex Gaussian, then Z is called a *complex Gaussian system*, or we simply call it *complex Gaussian.*

If Λ is the set of integers or an interval of the real line, we call Z a *complex Gaussian process.*

REMARK. If it is necessary to discriminate the system from a Gaussian system X introduced in Definition 2.1. in §1, we often call X a *real* Gaussian system.

Before we state some properties of a complex Gaussian system, it may be better to mention the reasons why such a complex system is of interest for us.

(i) When the spectral decomposition (to be discussed in the next chapter) of a stationary Gaussian process, it is necessary to form a random measure which is complex valued and which has its own meaning within electrical engineering as a random amplitude.

(ii) Once a given Gaussian process is complexified (to be a complex Gaussian system), then the Fourier transform is carried out within the framework of the system.

(iii) As will be clarified later, the linear structure of a Gaussian process can naturally be extended to that of a complex Gaussian process, and indeed, this structure is an important characteristic. Definition 2.2 is therefore quite natural.

(iv) In the analysis of nonlinear functionals of a Gaussian system (not discussed in this book) we often have variables of those functionals complexified. As a result, we are given a complex Gaussian system and we establish the analysis of functionals of a complex white noise.

We are now in a position to illustrate some basic properties of a complex Gaussian system. For the moment, we will assume that each random variable is centered in order to avoid unnecessary care. In fact, this assumption will never be essential.

Some easy consequences of the definition are now in order.

(1°) If a system $Z = \{Z_\lambda(\omega)\,;\ \lambda \in \Lambda\}$ is complex Gaussian, then the system $\{X_\lambda(\omega), Y_\lambda(\omega)\,;\ \lambda \in \Lambda\}$ of real random variables derived from the decomposition $Z_\lambda = X_\lambda + iY_\lambda$ is a real Gaussian system.

(2°) If $Z = \{Z_\lambda(\omega)\,;\ \lambda \in \Lambda\}$ is a system of complex Gaussian random variables which are mutually independent, then Z is complex Gaussian.

(3°) Given a complex Gaussian system $Z = \{Z_\lambda(\omega)\,;\ \lambda \in \Lambda\}$, we form a new system $\overline{Z} = \{\overline{Z}_\lambda(\omega)\,;\ \lambda \in \Lambda\}$, $\overline{Z}_\lambda$ being the complex conjugate of Z_λ. Then $\overline{Z}$ is also complex Gaussian. Except in the trivial case, $Z \cup \overline{Z}$ is not complex Gaussian.

(4°) Any subsystem of a complex Gaussian system Z is again complex Gaussian. If we add limits (in the sense of almost sure convergence, or of limit in the mean, or limit in probability) of members in Z to the original system Z, then we again have a complex Gaussian system (c.f. Theorem 2.2). The following proposition also shows that our definition of a complex Gaussian system is reasonable.

PROPOSITION 2.1. *Let $\{Z_1, Z_2\}$ be a complex Gaussian system. A necessary and sufficient condition for Z_1 and Z_2 to be independent is that their covariance vanishes.*

PROOF. As was assumed, Z_1 and Z_2 are centered. If they are independent, then the covariance $E(Z_1\overline{Z}_2)$ is equal to $E(Z_1)\cdot E(\overline{Z}_2) = 0$.

Conversely, if the covariance of Z_1 and Z_2 vanishes, we can prove independence in the following manner. Express Z_1 and Z_2 as in (2.9) with $m = 0$:

$$Z_j = X_j + iY_j, \qquad j = 1, 2.$$

By assumption

$$E\{(X_1 + iY_1)(X_2 - iY_2)\} = 0.$$

Hence we have

$$E(X_1X_2) + E(Y_1Y_2) = 0, \qquad E(Y_1X_2) - E(X_1Y_2) = 0.$$

On the other hand, the linear combinations $Z_1 + Z_2$ and $Z_1 + iZ_2$ are, by definition, both complex Gaussian. Since the real part and the imaginary part are mutually independent, we must have

$$E(Y_1X_2) + E(Y_1Y_2) = 0, \qquad E(X_1X_2) - E(Y_1Y_2) = 0.$$

Finally, we obtain

$$E(X_1X_2) = E(Y_1Y_2) = E(Y_1X_2) = E(X_1Y_2) = 0.$$

That is, $\{X_1, X_2, Y_1, Y_2\}$ is an independent system. In particular, the two random vectors (X_1, Y_1) and (X_2, Y_2) are independent, that is, Z_1 and Z_2 are independent. □

Associated with a general complex Gaussian system $Z = \{Z_\lambda(\omega)\,;\ \lambda \in \Lambda\}$ are the mean vector $m = (m_\lambda)$:

$$E(Z_\lambda) = m_\lambda, \qquad \lambda \in \Lambda,$$

and the covariance matrix $V = (V_{\lambda,\mu})$:

$$E\{(Z_\lambda - m_\lambda)(\overline{Z}_\mu - \overline{m}_\mu)\} = V_{\lambda,\mu}, \qquad \lambda, \mu \in \Lambda,$$

which is positive definite. The situation is the same as in the case of a real Gaussian system.

Now set

$$V_{\lambda,\mu} = 2(v_{\lambda,\mu} + iw_{\lambda,\mu}), \qquad v_{\lambda,\mu}, w_{\lambda,\mu} \text{ real}, \tag{2.10}$$

and set $Z_\lambda = X_\lambda + iY_\lambda + m_\lambda$ and $Z_\mu = X_\mu + iY_\mu + m_\mu$ as before. Noting that Z_λ, Z_μ and $Z_\lambda + iZ_\mu$ are all complex Gaussian, we can prove the following:

$$v_{\lambda,\mu} = v_{\mu,\lambda} = E(X_\lambda X_\mu) = E(Y_\lambda Y_\mu),$$
$$w_{\lambda,\mu} = -w_{\mu,\lambda} = E(Y_\lambda X_\mu) = -E(X_\lambda Y_\mu).$$

The matrices with entries $v_{\lambda,\mu}$ and $w_{\lambda,\mu}$ are denoted by v and w, respectively:

$$v = (v_{\lambda,\mu}), \qquad w = (w_{\lambda,\mu}). \tag{2.11}$$

Lemma. *Let $V = (V_{\lambda,\mu})$ be a positive definite matrix with $V_{\lambda,\mu}$ complex. Take the real matrices v and w determined by* (2.10) *and* (2.11) *and define a new matrix $\mathfrak{v}$ by*

$$\mathfrak{v} = \begin{pmatrix} v & -w \\ w & v \end{pmatrix}. \tag{2.12}$$

Then $\mathfrak{v}$ is also positive definite.

Proof. Take finite λ-set $\{\lambda\}$, and let $\{\alpha_\lambda\}$ and $\{\beta_\lambda\}$ be any finite sets of complex numbers. We form

$$Q = \sum_{\lambda,\mu} \alpha_\lambda \overline{\alpha}_\mu v_{\lambda,\mu} + \sum_{\lambda,\mu} \beta_\lambda \overline{\beta}_\mu v_{\lambda,\mu} + \sum_{\lambda,\mu} \beta_\lambda \overline{\alpha}_\mu w_{\lambda,\mu} - \sum_{\lambda,\mu} \alpha_\lambda \overline{\beta}_\mu w_{\lambda,\mu}.$$

Since $(V_{\lambda,\mu})$ is positive definite, so is $(\overline{V}_{\lambda,\mu})$. This implies

$$\sum_{\lambda,\mu}(\alpha_\lambda - i\beta_\lambda)(\overline{\alpha_\mu - i\beta_\mu})(v_{\lambda,\mu} + iw_{\lambda,\mu}) \geq 0,$$
$$\sum_{\lambda,\mu}(\alpha_\lambda + i\beta_\lambda)(\overline{\alpha_\mu + i\beta_\mu})(v_{\lambda,\mu} - iw_{\lambda,\mu}) \geq 0.$$

Therefore, we have

$$\sum_{\lambda,\mu}\{(\alpha_\lambda - i\beta_\lambda)(\overline{\alpha}_\mu + i\overline{\beta}_\mu) + (\alpha_\lambda + i\beta_\lambda)(\overline{\alpha}_\mu - i\overline{\beta}_\mu)\}v_{\lambda,\mu}$$
$$+ \sum_{\lambda,\mu}\{i(\alpha_\lambda - i\beta_\lambda)(\overline{\alpha}_\mu + i\overline{\beta}_\mu) - i(\alpha_\lambda + i\beta_\lambda)(\overline{\alpha}_\mu - i\overline{\beta}_\mu)\}w_{\lambda,\mu} \geq 0.$$

This is the same as

$$2\sum_{\lambda,\mu}(\alpha_\lambda\overline{\alpha}_\mu + \beta_\lambda\overline{\beta}_\mu)v_{\lambda,\mu} + 2\sum_{\lambda,\mu}(\beta_\lambda\overline{\alpha}_\mu - \alpha_\lambda\overline{\beta}_\mu)w_{\lambda,\mu} \geq 0,$$

which shows that $Q \geq 0$. Hence, $\mathfrak{v}$ is positive definite. □

We are now ready to state the main theorem in this section.

THEOREM 2.4. *Take Λ to be a parameter set. For a given vector $m = (m_\lambda)$ and a positive definite matrix $V = (V_{\lambda,\mu})$, there exists a complex Gaussian system $Z = \{Z_\lambda(\omega)\,;\ \lambda \in \Lambda\}$ with mean vector m and covariance matrix V.*

PROOF. We form a real positive definite matrix $\mathfrak{v}$, by the above lemma, from the V that is given. Then by Theorem 2.1, we can form a (real) Gaussian system $\{X_\lambda, Y_\lambda\,;\ \lambda \in \Lambda\}$ with mean vector 0 and covariance matrix $\mathfrak{v}$, where both $\{X_\lambda\}$ and $\{Y_\lambda\}$ have the same covariance matrix v and where

$$E(Y_\lambda X_\mu) = -E(X_\lambda Y_\mu) = w_{\lambda,\mu}.$$

Using the given vector m, we define

$$Z_\lambda(\omega) = X_\lambda(\omega) + iY_\lambda(\omega) + m_\lambda, \qquad \lambda \in \Lambda.$$

It is easy to see that $\{Z_\lambda(\omega)\,;\ \lambda \in \Lambda\}$ is a complex Gaussian system. □

Here is a short remark. If in the above theorem, all the entries of the given matrix V are real, the theorem is still, of course, true. It is, however, seen that

$$w_{\lambda,\mu} = 0 \quad \text{for every } \lambda,\ \mu$$

by (2.10). Hence, the two systems $\{X_\lambda\}$ and $\{Y_\lambda\}$ appearing in the proof are independent.

The above theorem can be applied to Gaussian processes with discrete or continuous parameter to prove the following proposition.

PROPOSITION 2.2. *Let a real Gaussian process be given. Then we can form a complex Gaussian process with the same covariance.*

PROOF. Let a Gaussian process $\{X_\lambda\}$ be given. As before, we may assume that $E(X_\lambda) = 0$ for every λ. Take a copy of $\{X_\lambda\}$ and denote it by $\{Y_\lambda\}$ which is assumed to be independent of $\{X_\lambda\}$. Set

$$Z_\lambda(\omega) = \frac{1}{\sqrt{2}}\{X_\lambda(\omega) + Y_\lambda(\omega)\} \tag{2.13}$$

Then $\{Z_\lambda\}$ is exactly what we want to have. □

DEFINITION 2.3. A complex Gaussian process which is formed as in formula (2.13) is called a *complex form* of $\{X_\lambda\}$.

It is instructive to consider some examples.

EXAMPLE 1. Take $\Lambda = [0, \infty)$, $m_\lambda \equiv 0$, and $v_{\lambda,\mu} = \min\{\lambda, \mu\}$, denoted by $\lambda \wedge \mu$. A Gaussian system determined by Theorem 2.1 with these choices of m_λ and $V_{\lambda,\mu}$ is called a *Brownian motion.* We prefer, in this case, the notation $\{B(t);\ t \in [0, \infty)\}$ rather than $\{X_\lambda\}$. A complex form of $\{B(t)\}$ is called a *complex Brownian motion* and is usually denoted by $\{Z(t)\}$.

EXAMPLE 2. Let $\Lambda = \{\lambda;\ \text{Borel set} \subset R,\ |\lambda| < \infty\}$, where $|\lambda|$ is the Lebesgue measure of λ. Take $m_\lambda \equiv 0$ and $V_{\lambda,\mu} = |\lambda \cap \mu|$. As is well known, $V = (V_{\lambda,\mu})$ is positive definite. With this choice of $m = (m_\lambda)$ and V, we can form a real Gaussian system, which is called a *homogeneous Gaussian random measure.* Its complex form, often denoted by $Z = \{Z(\lambda);\ \lambda \in \Lambda\}$, is a *complex homogeneous Gaussian random measure.*

§4. Discrete parameter Gaussian processes: Canonical representation

A real Gaussian system $X = \{X_\lambda(\omega);\ \lambda \in \Lambda\}$ is said to be

(i) a *discrete parameter* Gaussian process if Λ is the set of whole integers I or the set I^+ of all positive integers, and

(ii) a *continuous parameter* Gaussian process if Λ is either the whole of R or an interval of R.

To emphasize these cases we use the notation X_n in the case (i), while we write $X(t)$ in the case (ii) instead of X_λ. The symbols n and t are supposed to stand for the time not simply as a parameter.

The difference between the cases (i) and (ii) is relatively large. We shall start with the study of the canonical representation of the process in the easier case, that is case (i).

Take a discrete parameter Gaussian process $X = \{X_n(\omega);\ n \in I^+\}$. Here $E(X_n) \equiv 0$ is still assumed. Since each X_n has finite variance, the X_n may be viewed as members of the (real) $L^2(\Omega, P)$ which is the Hilbert space involving all real random variables with finite variance. We can therefore apply the Schmidt orthogonalization technique to the X_n, $n = 1, 2, \dots$, to obtain

$$X_1 = a_{1,1}\xi_1,$$
$$X_2 = a_{2,1}\xi_1 + a_{2,2}\xi_2,$$
$$\cdots\cdots$$

In general,

$$X_n = \sum_{j \le n} a_{n,j}\xi_j, \tag{2.14}$$

where the $a_{n,j}$, $j \le n$, are all real and where $\{\xi_n\}$ is an orthonormal basis in $L^2(\Omega, P)$. The method used to form the ξ_j shows that each ξ_j is a linear combination of the X_k, $k \le j$. Since $\{X_n\}$ is Gaussian, so is $\{\xi_n; n \in I^+\}$. Hence the ξ_n's are mutually independent. (See §2, (4°).)

As for the conditional expectation, we have an expression of the form

$$E(X_n|X_1, X_2, \ldots, X_k) = \sum_{j=1}^{k} a_{n,j}\xi_j, \qquad k < n. \tag{2.15}$$

Because X_n is expanded in terms of the orthonormal basis $\{\xi_j\}$ and the sum is divided into two parts,

$$X_n = \sum_{j=1}^{k} a_{n,j}\xi_j + \sum_{j=k+1}^{n} a_{n,j}\xi_j.$$

The first sum is a linear function of X_j, $1 \le j \le k$, while the second term is independent of those X_j's for $j \le k$. This proves (2.15).

Setting $k = n - 1$, we obtain

$$X_n - E(X_n|X_1, X_2, \ldots, X_{n-1}) = a_{n,n}\xi_n. \tag{2.16}$$

If $a_{n,n} \ne 0$, then ξ_n is thought of as a random variable that contains the full, new marginal information gained by X_n at the instant n. In addition, ξ_n is independent of all the X_j, $j \le n-1$, and X_n itself is expressed as a function of ξ_n and X_j, $j \le n-1$.

The above discussion illustrates that the representation (2.14) of X_n in terms of the kernel $a_{n,j}$ and ξ_j is fitting for our purpose to observe the development of the random phenomena described by X_n, as time n goes by. Also we can see the importance of the roles played by the ξ_n from the view point of the information theory. Thus we are led to give

DEFINITION 2.4. Let $X = \{X_n(\omega); n \in I^+\}$ be a Gaussian process with $E(X_n) = 0$ for every n.

(i) If there exists a double sequence $a_{n,j}$, $j \le n$, such that for a system of independent standard Gaussian random variables $\{\xi_n; n \in I^+\}$ the process X and $Y = \{Y_n; n \in I^+\}$ with

$$Y_n = \sum_{j \le n} a_{n,j}\xi_j$$

have the same probability distribution, then the system $\{a_{n,j},\ j \le n;\ \xi_n,\ n \in I^+\}$ (or denoted by $\{a_{n,j},\ \xi_n\}$) is called a *representation* of X. Sometimes the above formula for Y_n is called a representation of X.

(ii) If a representation $\{a_{n,j},\ \xi_n\}$ satisfies conditions (2.14) and (2.15), then it is called a *canonical representation.*

(iii) The $\{\xi_n\}$ of a canonical representation $\{a_{n,j},\ \xi_n\}$ is called the *innovation process* or simply the *innovation* of the process X; and the $a_{n,j}$ is called the *canonical kernel* of the representation.

We note that there are, in general, many noncanonical representations, which are of course not interesting and not useful in actual applications. The following example illustrates how noncanonical representations can appear.

EXAMPLE 1. Assume that $\{X_1, X_2, X_3\}$ is Gaussian and that the X_i are expressed in the form

$$X_1 = a_{1,1}\xi_1, \quad X_2 = ca_{1,1}\xi_1 + 0\xi_2, \quad X_3 = a_{3,1}\xi_1 + 0\xi_2 + a_{3,3}\xi_3, \qquad (*)$$

where ξ_1, ξ_2, ξ_3 are independent $N(0, 1)$ variables. Another system, denoted by $\{\tilde{X}_1, \tilde{X}_2, \tilde{X}_3\}$, is defined by

$$\tilde{X}_1 = X_1, \quad \tilde{X}_2 = X_2, \quad \tilde{X}_3 = a_{3,1}\xi_1 + (a_{3,3}/\sqrt{2})\xi_2 + (a_{3,3}/\sqrt{2})\xi_3. \qquad (**)$$

Obviously the two systems $\{X_1, X_2, X_3\}$ and $\{\tilde{X}_1, \tilde{X}_2, \tilde{X}_3\}$ have the same probability distribution. But the latter does not satisfy (2.15). In fact, we have

$$E(X_3|X_1, X_2) = a_{3,1}\xi_1,$$

that is, a canonical representation is given by $(*)$. The ξ_2 is in fact unnecessary for the representation and $\{\xi_1, 0, \xi_3\}$ is the innovation. While

$$E(\tilde{X}_3|\tilde{X}_1, \tilde{X}_2) \ne a_{3,1}\xi_1 + (a_{3,3}/\sqrt{2})\xi_2,$$

so that $(**)$ is not a canonical representation of $\{X_1, X_2, X_3\}$,

THEOREM 2.5. *A Gaussian process $X = \{X_n(\omega);\ n \in I^+\}$ always has a canonical representation. It is unique in the sense that if both the representations $\{a_{n,j}, \xi_n\}$ and $\{a'_{n,j}, \xi'_n\}$ are canonical, then for every n, either*

$$a_{m,n} = a'_{m,n} \quad \textit{for all } m \ge n$$

or

$$a_{m,n} = -a'_{m,n} \quad \textit{for all } m \ge n.$$

PROOF. The existence of a canonical representation is guaranteed by Schmidt's orthogonalization theorem, as we have seen at the beginning of this section.

As for the uniqueness, we first conclude $a^2_{n,n} = a'^2_{n,n}$ since the left-hand side of (2.16) is independent of the choice of a representation. Once we choose the case $a_{n,n} = a'_{n,n}$, the covariance of X_m and X_n $(m < n)$

$$E(X_m X_n) = \sum_{j=1}^{m} a_{m,j} a_{n,j} = \sum_{j=1}^{m} a'_{m,j} a'_{n,j} \tag{2.17}$$

uniquely determines $a_{n,j} = a'_{n,j}$ successively in j $(\leq n)$. Another choice $a_{n,n} = -a'_{n,n}$ is possible, and we can play exactly the same game to arrive at the conclusion of the theorem. □

Roughly speaking, the theorem says that in the determination of the canonical kernel there is freedom only to choose the sign of $a_{n,n}$.

The situation is exactly the same for a complex Gaussian process $Z = \{Z_n(\omega);\ n \in I^+\}$. Namely, we always find the canonical representation in such a way that there exists a sequence $\{\xi_n\}$ of independent complex Gaussian random variables with $E(\xi_n) = 0$ and $E(|\xi_n|^2) = 1$ and such that each Z_n can be expressed in the form

$$Z_n = \sum_{j=1}^{n} a_{n,j} \xi_j, \qquad a_{n,j} \in C \tag{2.18}$$

satisfying the equation

$$E(Z_n | Z_1, Z_2, \dots, Z_k) = \sum_{j=1}^{k} a_{n,j} \xi_j, \qquad k \leq n. \tag{2.19}$$

As in the case of (real) Gaussian process, the system $\{a_{n,j},\ j \leq n;\ \xi_n,\ n \in I^+\}$ is called the *canonical representation* of Z and $\{\xi_n;\ n \in I^+\}$ is said to be the *innovation process* of Z. As for $a_{n,n}$, we have the freedom to multiply by a factor of unit modulus.

Before we proceed to the next topic, we illustrate with an example.

EXAMPLE 2. Let $\{a_{n,j}, \xi_n\}$ be the canonical representation of a real Gaussian process $X = \{X_n(\omega);\ n \in I^+\}$. Assume that the canonical kernel $a_{n,j}$ is of the following particular form

$$a_{n,j} = ac^{n-j}, \qquad j \leq n.$$

Then for $n > k$, the conditional expectation has the form

$$E(X_n | X_1, X_2, \dots, X_k) = \sum_{j=1}^{k} ac^{n-j} \xi_j = c^{n-k} X_k.$$

Hence, X_n is expressible as a sum of two independent terms

$$X_n = c^{n-k} X_k + \sum_{j=k+1}^{n} ac^{n-j} \xi_j.$$

This implies that the remarkable relation

$$P(X_n \leq x | X_1, X_2, \dots, X_k) = P(X_n \leq x | X_k), \qquad x \in R,$$

is valid for every pair (n, k) with $k \le n$.

Here is a somewhat general definition.

DEFINITION 2.5. A stochastic process $X = \{X_n(\omega); n \in I^+\}$ is a *Markov process* if for any n, k $(k < n)$

$$P(X_n \le x | X_1, X_2, \dots, X_k) = P(X_n \le x | X_k), \qquad x \in R$$

holds a.e. P.

The Gaussian process in Example 2 is a Markov process.

We have a condition on a Gaussian process X that is nondegenerate in the sense that $E(X_n | X_k, k \le n-1) \ne X_n$. In fact, the canonical kernel gives us a condition for X to be a Markov process.

THEOREM 2.6. *A necessary and sufficient condition for a nondegenerate Gaussian process $X = \{X_n(\omega); n \in I^+\}$ to be a Markov process is that the canonical kernel has the form*

$$a_{n,j} = a_n \cdot b_j, \qquad j \le n. \tag{2.20}$$

PROOF. *Sufficiency.* Suppose that X_n is expressed, using the kernel $a_{n,j} = a_n \cdot b_j$, in the form

$$X_n = \sum_{j=1}^{n} a_n \cdot b_j \xi_j.$$

Then for any pair (n, k) with $k \le n$ we have

$$E(X_n | X_1, X_2, \dots, X_k) = \sum_{j=1}^{k} a_n \cdot b_j \xi_j = \frac{a_n}{a_k} X_k \quad \text{a.e. } P.$$

Hence, X_n admits a decomposition similar to that in Example 2:

$$X_n = \frac{a_n}{a_k} X_k + \sum_{j=k+1}^{n} a_n \cdot b_j \xi_j$$

and we prove that X is a Markov process.

Necessity. If X is a Markov process,

$$E(X_n | X_1, X_2, \dots, X_k) = E(X_n | X_k) \quad \text{a.e. } P$$

holds for any pair (n, k) with $k \le n$. Let $\{a_{n,j}, \xi_n\}$ be the canonical representation. Then the left-hand side of the above formula is $\sum_{j=1}^{k} a_{n,j} \xi_j$. Since $\{X_n, X_k\}$ is Gaussian, the right-hand side is a linear function of X_k, that is const$\cdot X_k$. Thus, we have

$$\begin{aligned} \sum_{j=1}^{k} a_{n,j} \xi_j &= a_{n,k} X_k \\ &= c_{n,k} \sum_{j=1}^{k} a_{k,j} \xi_j, \qquad c_{n,k} \text{ constant}. \end{aligned}$$

The ξ_j, $1 \le j \le k$, are linearly independent and the canonical kernel is unique up to sign of the $a_{n,n}$, so that we must have the $a_{n,j}$, $n > j$, expressed in the form

$$a_{n,j} = c_{n,k} \cdot a_{k,j}, \qquad 1 \le j \le k \le n.$$

Since k is arbitrary, equation (2.20) is established. □

REMARK. The determination (2.20) of the canonical kernel never violates the uniqueness of the kernel stated in the above theorem.

The Markov property of a Gaussian process $X = \{X_n;\, n \in I^+\}$ can be characterized in terms of the covariance function, using the above theorem.

COROLLARY. *A nondegenerate Gaussian process is Markov if and only if the entries* $\Gamma_{m,n}$ *of its covariance matrix* Γ *are expressed in the form*

$$\Gamma_{m,n} = a_m a_n c_n, \qquad m \ge n, \tag{2.21}$$

where $a_n \ne 0$ *and* c_n (> 0) *is increasing in* n.

PROOF. We may assume $E(X_n) = 0$ for every n. If X is a Markov process, then (2.20) implies

$$E(X_m X_n) = \sum_{j=1}^{n} (a_m b_j)(a_n b_j) = a_m a_n \sum_{j=1}^{n} b_j^2,$$

so that by setting

$$c_n = \sum_{j=1}^{n} b_j^2 \tag{2.22}$$

we obtain formula (2.21).

The converse is proved as follows. Let $\Gamma_{m,n}$ be given as in the formula (2.21) with $c_n = \sum_{j=1}^{n} b_j^2$, the sign of b_j being arbitrary. We then form $Y = \{Y_n;\ n \in I^+\}$ by

$$Y_n = a_n \sum_{j=1}^{n} b_j \xi_j,$$

where $\{\xi_n\}$ is an independent system of standard Gaussian random variables. The covariance matrix of Y is in agreement with (2.21). Hence X and Y are the same Gaussian process. To show the Markov property of Y, it suffices to prove that $\{a_n b_j, \xi_n\}$ is the canonical representation (Theorem 2.6.). Noting $a_n b_n \ne 0$, it can be proved that the two systems $\{Y_1, Y_2, \dots, Y_n\}$ and $\{\xi_1, \xi_2, \dots, \xi_n\}$ span the same vector space. This implies (2.15), which is what was to be proved. □

What has been discovered so far is that the Gaussian Markov property can clearly be characterized in terms of the canonical kernel. The next step might be a generalization of this fact; namely a characterization of a multiple Markov Gaussian process, although we have to think seriously of how to define the multiple Markov property before we return to the present question. This property will be discussed in Chapter V.

§5. Continuous parameter Gaussian processes: Brownian motion

As was clarified in the last section, a system $\xi = \{\xi_n ; n \in I^+\}$ is an important and basic process by which general Gaussian processes can be formed; in particular, an innovation process is of this type, being subordinated to the given process. A continuous analogue of ξ_n may be taken to be the increment of a Brownian motion $B = \{B(t) ; t \in [0, \infty)\}$ introduced in Example 1, §3. We shall see in Theorem 2.7 the reason why this is so . Furthermore, if one wishes to find a formal analogue of ξ, then it could be a *white noise* $\dot{B} = \{\dot{B}(t) ; t \in [0, \infty)\}$ with $\dot{B}(t) = \frac{d}{dt}B(t)$, which has independent values at every point in time. Unfortunately, $\dot{B}$ is a generalized process in the sense that the sample functions are generalized functions. We therefore prefer B to $\dot{B}$, and we often use its infinitesimal increment $dB(t)$ if necessary.

Since the roles of ξ in the representation theory are played by a Brownian motion in the continuous parameter case, we now present some basic properties of a Brownian motion as background for representation theory. The following theorem illustrates the independent increments property of a Brownian motion.

THEOREM 2.7. *Let $B = \{B(t) ; t \in [0, \infty)\}$ be a Brownian motion. For any nonoverlapping intervals $[s_i, t_i]$, $i = 1, 2, \dots, n$, the family*

$$\Xi = \{B(t_i) - B(s_i) ; \ i = 1, 2, \dots, n\}$$

forms an independent system.

PROOF. The covariance matrix $V = (V_{i,j})$ has the form

$$\begin{aligned} V_{i,i} &= t_i - s_i, \\ V_{i,j} &= E[(B(t_i) - B(s_i))(B(t_j) - B(s_j))] \\ &= t_i - s_i + t_i - s_i = 0, \qquad s_i < t_i \le s_j < t_j . \end{aligned}$$

Hence, by (5) in §2, Ξ is an independent system. □

Denote by $\mathbf{B}_t(B)$ the smallest σ-field with respect to which all the $B(s)$, $s \le t$, are measurable ($\mathbf{B}_t(B)$ is the σ-field generated by the $B(s)$, $s \le t$).

COROLLARY. *Take the $\mathbf{B}_t(B)$. Then $\{B(t), \mathbf{B}_t(B), t \in [0, \infty)\}$ is a martingale.*

PROOF. We have to prove

$$E(B(t)|\mathbf{B}_s(B)) = B(s), \qquad s < t, \quad \text{a.e. } P. \tag{2.23}$$

By Theorem 2.7, $B(t) - B(s)$ is independent of all the $B(u) - B(v)$, $0 \le u, v \le s$ $(\le t)$, in particular, independent of $B(u) = B(u) - B(0)$, for all $u \le s$. Hence, $E(B(t) - B(s)|\mathbf{B}_s(B)) = 0$, a.e. P. This implies (2.23). □

REMARK. We often take a sequence of σ-fields $\mathbf{F}_t$ (that is wider than $\mathbf{B}_t(B)$) with respect to which B is still a martingale. Choose, for instance, a σ-field $\mathbf{G}$ independent of $\mathbf{B}_\infty(B)$ and set

$$\mathbf{F}_t = \mathbf{G} \vee \mathbf{B}_t(B). \qquad (\vee \text{ denotes the lattice sum.})$$

Then $\{B(t), \mathbf{F}_t, t \in [0, \infty)\}$ is again a martingale.

Some sample function (or path) properties of a Brownian motion are now briefly mentioned.

THEOREM 2.8. *Almost all sample functions* $B(\cdot, \omega)$ *are continuous.*

This assertion may be rephrased as follows: Let μ^W be the probability distribution of a Brownian motion B over $R^{[0,\infty)}$, and let $C = C([0, \infty))$ be the set of all continuous functions on $[0, \infty)$. Then

$$\tilde{\mu}^W(C) = 1,$$

where $\tilde{\mu}^W$ is the outer measure derived from μ^W.

PROOF OF THEOREM 2.8. Computation of the fourth order moment of a Gaussian distribution tells us that

$$E[|B(t) - B(s)|^4] = 3|t - s|^2.$$

Hence, the Kolmogorov-Prokhorov criterion (Theorem 1.14) guarantees the result. □

REMARK. The probability distribution of a Brownian motion can now be considered as a probability measure on C (cf. §1.6).

DEFINITION 2.6. The probability measure μ^W on C derived from a Brownian motion is called a *Wiener measure* (N. Wiener (1923)), and the measure space $(C, \mathbf{C}, \mu^W)$ is called the *Wiener space*, where $\mathbf{C}$ is the σ-field generated by the open subsets of C.

Beside the continuity of sample functions, their variations are worth mentioning.

THEOREM 2.9. *Let* Δ *be a partition of the time interval* $[s, t]$.

$$\Delta : s = t_0 < t_1 < \cdots < t_n = t.$$

We set

$$\chi_\Delta = \sum_{i=1}^{n} (B(t_i) - B(t_{i-1}))^2.$$

Then the χ_Δ *converges to* $t - s$ *in probability as* $\delta(\Delta)$ *tends to* 0, *where* $\delta(\Delta) = \max |t_i - t_{i-1}|$.

PROOF. Note that

$$E(\chi_\Delta) = t - s \quad \text{and} \quad V(\chi_\Delta) = 2\sum_{i=1}^{n} (t_i - t_{i-1})^2 \le 2(t - s)\delta(\Delta).$$

The assertion is then easily proved by the Tchebychev inequality (Proposition 1.2 (1°)). □

COROLLARY. *For almost all ω, the sample function $B(\cdot, \omega)$ of a Brownian motion is of unbounded variation on any time interval.*

PROOF. Suppose that a sample function $B(t) = B(t, \omega)$ is of bounded variation on an interval $[s, t]$. Then for any partition Δ of this interval $\sum_{i=1}^{n} |B(t_i) - B(t_{i-1})|$ is bounded (say, $\leq K = K(\omega)$). Then we must have

$$\chi_\Delta = \sum_{i=1}^{n} |B(t_i) - B(t_{i-1})|^2 \leq K \cdot \max_{1\leq i\leq n} |B(t_i) - B(t_{i-1})|.$$

Let $\delta(\Delta)$ tend to 0. Then the right-hand side converges to 0 by the uniform continuity of a sample function. So does the variation χ_Δ. This contradicts Theorem 2.9. □

Our next goal is to define an integral with respect to $dB(t)$ in order to get a representation of a Gaussian process in terms of a Brownian motion. According to Theorem 2.9 we are not permitted to define an integral based on $dB(t)$ as a Lebesgue-Stieltjes integral. In spite of this fact, it is possible to define a *stochastic integral* based on dB, as is explained below. We restrict our attention to integrals with nonrandom integrand.

Let $f \in L^2([0, \infty))$ be a step function:

$$f(t) = \sum_{i=1}^{n} a_i \chi_{I_i}(t), \qquad I_i = [t_{i-1}, t_i),$$

where $t_0 = 0 < t_1 < t_2 < \cdots < t_n = \infty$, and χ_I denotes the indicator function of the set I. For such a function f we define

$$\int_0^\infty f(t)\, dB(t) = \sum_{i=1}^{n} a_i (B(t_i) - B(t_{i-1})).$$

It is easy to prove

LEMMA. *For a simple function $f \in L^2([0, \infty))$, $\int_0^\infty f(t)\, dB(t)$ is a centered Gaussian random variable with variance*

$$E\left(\left|\int_0^\infty f(t)\, dB(t)\right|^2\right) = \int_0^\infty f(t)^2 dt = \|f\|^2.$$

This lemma suggests to us that we define the integral for a general f in $L^2([0, \infty))$ in such a way that

$$\int_0^\infty f(t)\, dB(t) = \operatorname*{l.\,i.\,m.}_{n\to\infty} \int_0^\infty f_n(t)\, dB(t) \tag{2.24}$$

(l. i. m. means the strong limit in $L^2(\Omega, P)$), by taking a sequence $\{f_n\}$ of step functions such that $\|f - f_n\| \to 0$ in $L^2([0, \infty))$. Note that such an integral is well defined (the limit is independent of the choice of a sequence f_n, and so forth) and understand that the formula (2.24) is a reasonable way of defining the integral based on dB. An integral $\int_0^\infty f(t)\, dB(t)$ thus defined is called a *Wiener integral*.

THEOREM 2.10. (1°) *Set* $\mathscr{H}_1 = \{\int_0^\infty f(t)\,dB(t)\,;\ f \in L^2([0,\infty))\}$ *the collection of Wiener integrals. Then* $\mathscr{H}_1$ *forms a Gaussian system with mean vector* 0 *and covariance matrix is*

$$V_{f,g} = E\left[\int_0^\infty f(t)\,dB(t) \int_0^\infty g(t)\,dB(t)\right] = (f, g), \qquad f, g \in L^2([0,\infty)),$$

where (,) *denotes the inner product in* $L^2([0,\infty))$.

(2°) *For* $f \in L^2([0,\infty))$ *we set* $\int_0^t f(s)\,dB(s) = \int_0^\infty \chi_{[0,t)}(s) f(s)\,dB(s)$. *Then* $\{\int_0^t f(s)\,dB(s),\ \mathbf{B}_t(B),\ t \in [0,\infty)\}$ *is a regular martingale.*

PROOF. (1°) is almost obvious by definition. As for (2°) we can prove this first for a step function f. Then for a general L^2-function f we can use approximations by step functions to prove the martingale equality. □

REMARK. The system $\mathscr{H}_1$ is equal to the closed linear hull $H(B)$ spanned by the Gaussian system $B = \{B(t)\,;\ t \in [0,\infty)\}$ (c.f. §1).

We are now ready to discuss canonical representations based on B. Once again we note that a counter part of the innovation process $\{\xi_n\}$ in the discrete parameter case is taken to be $dB(t)$. For simplicity, we assume that all the Gaussian processes to be discussed are centered as before.

DEFINITION 2.7. Let $X = \{X(t)\,;\ t \in [0,\infty)\}$ be a Gaussian process. If there exists a kernel function $F(t, u)$ such that $Y = \{Y(t)\,;\ t \in [0,\infty)\}$ given by

$$Y(t) = \int_0^t F(t, u)\,dB(u) \quad \text{(Wiener integral)} \tag{2.25}$$

is the same process as X and if $\mathbf{B}_t(Y) = \mathbf{B}_t(B)$ holds for every t, then $\{F(t, u),\ u \le t\,;\ B(t)\}$ (simply written as $\{F(t, u), B\}$) is called the *canonical representation* of X, and $\{F(t, u)$ is the *canonical kernel*.

The $X(t)$ above is often said to be *represented canonically* in the form (2.25).

There are cases where $X(t)$ itself is expressed as in (2.25); $Y(t) = X(t)$. Then, the Brownian motion B above is the innovation process of X, and the integral (2.25) is often called the canonical representation of X.

There are of course many representations of X of the form (2.25) which are not canonical. Unlike the discrete parameter case, the existence of the canonical representation is not always guaranteed for continuous parameter Gaussian processes. This situation will be clarified when the general theory of the canonical representation is discussed in Chapter IV. Here we present an example for which the canonical representation is easily formed.

EXAMPLE. A Gaussian process $X = \{X(t)\,;\ t \in [0,\infty)\}$ given by

$$X(t) = F(t) \int_0^t G(u)\,dB(u), \tag{2.26}$$
$$F, G : \text{continuous}, \qquad F(t) \neq 0,\ G(t) \neq 0 \quad \text{for } t > 0$$

is represented canonically by a Brownian motion B. In fact, $\mathbf{B}_t(X) \subset \mathbf{B}_t(B)$ is obvious. The converse inclusion is proved in the following manner. Let

$H_t(B)$ be the closed linear hull spanned by the Wiener integrals of the form $\int_0^t f(u)\,dB(u)$, $f \in L^2([0, \infty))$. If $\mathbf{B}_t(X)$ is a proper subclass of $\mathbf{B}_t(B)$, then there exists an element Y in $H_t(B)$ such that $Y = \int_0^t f(u)\,dB(u) \neq 0$ and such that $E(X(s)Y) = 0$ for all $s \leq t$.

Hence, we have for any $s \leq t$,

$$E\left\{F(s)\int_0^s G(u)\,dB(u)\cdot\int_0^t f(u)\,dB(u)\right\} = F(s)\int_0^s G(u)f(u)\,du = 0.$$

Since $F(s) \neq 0$, we must have $f(u) = 0$, a.e., which leads us to the contradiction that $Y \neq 0$. Hence, $\mathbf{B}_t(X) = \mathbf{B}_t(B)$ is true for any t, that is, (2.26) gives a canonical representation.

The Markov property comes from

$$E(X(t)|\mathbf{B}_s(X)) = E(X(t)|X(s)), \qquad s \leq t, \tag{2.27}$$

which is easily proved by (2.26) and by the property $\mathbf{B}_t(X) = \mathbf{B}_t(B)$.

If a given Gaussian process X satisfies the Markov property, then (2.27) holds. With this assumption we can arrive essentially at the expression (2.26). This is a counterpart to Theorem 2.6 for the discrete parameter case. Details will be discussed in a more general setting in Chapter V. Roughly speaking, an innovation process $B = \{B(t)\}$ is slightly more general, and its covariance function (matrix) is given by the formula

$$E(B(t)B(s)) = \int_0^{t\wedge s} m(du), \qquad m \text{ a continuous measure}. \tag{2.28}$$

$B(t)$ has independent increments and, with this property, (general) stochastic integrals can be defined for an integrand f in the space $L^2(m) = \{f;\ \int_0^\infty f(u)^2\,m(du) < \infty\}$.

CHAPTER III

Stationary Gaussian Processes and Their Representations

A mean-square continuous stationary Gaussian process $X = \{X(t); t \in R\}$, with constant mean and with covariance function $\Gamma(t, s) = \gamma(t-s)$ depending only on $t-s$, is an important object in the study of stochastic processes. Associated with a stationary Gaussian process is a shift operator $T_t: X(s) \dashrightarrow X(s+t)$, which is a unitary operator acting on the Hilbert space $H(X)$ spanned by the $X(t)$, and the T_t form a one-parameter group: $T_t T_s = T_{t+s}$, for t, $s \in R$. We can therefore discuss the spectral resolution of T_t, and by using the spectral measure, we can form the moving average representation of the given process as well as the prediction operator. Following this result, we discuss the representation theory of stationary Gaussian processes. The basic idea of canonical representation was presented in the last chapter, and now we are given a stationary process for which the spectral measure is available, as additional information to form the canonical kernel in an analytic manner. With this information the canonical representation of the given Gaussian process in terms of the Brownian motion can be formed systematically. The method for constructing the representation will be clarified in the discrete parameter case in §3. It will be shown in §4 that the method can be applied to the continuous parameter case in a natural manner. There one should note that, since we are assuming the processes to be Gaussian, the mean and the covariance function provide enough information to determine the distribution of the process X. Note also that the distribution is invariant under the time shift acting on the individual sample function, that is, X is strictly stationary.

§1. Discrete parameter stationary processes

We first define a stationary (not necessarily Gaussian) process with discrete parameter. Since the time variable n shall be shifted back and forth freely, the time parameter set has to be the set I of the whole integers.

Assume that a stochastic process $X = \{X_n(\omega); \ n \in I\}$ is given on a probability space $(\Omega, \mathbf{B}, P)$.

DEFINITION 3.1. If, for any integers n and h, two random vectors

$$(X_n, X_{n+1}, \ldots, X_{n+k-1}) \quad \text{and} \quad (X_{n+h}, X_{n+h+1}, \ldots, X_{n+h+k-1})$$

have the same probability distribution, then X is said to be a *stationary process.*

If, in addition, X is Gaussian, then it is called a *stationary Gaussian process.*

Returning to a general stationary process X, we now assume that the X_n have moments up to the second order. Set $k = 1$ in the above definition. Then, it is seen that each X_n has the same distribution. In particular, we have

$$E(X_n) = m, \qquad m \text{ being a constant independent of } n. \tag{3.1}$$

Also, by definition, the joint distribution of (X_n, X_{n+h}) does not depend on n, so neither does the covariance function

$$E\{(X_{n+h} - m)(X_n - m)\} = \gamma_h. \tag{3.2}$$

Since the X_n is real valued, we claim

$$\gamma_{-h} = \gamma_h \quad \text{for every } h \in I. \tag{3.3}$$

The above three conditions (3.1–3.3) are all necessary for X to be stationary under the assumption of the existence of the first and the second moments of the X_n. With this in mind we are going to define a weakly stationary process.

DEFINITION 3.2. Let $X = \{X_n(\omega);\ n \in I\}$ be a process such that each X_n has finite moments up to the second order and that the mean and covariance function satisfy (3.1) and (3.2), respectively. Then X is said to be *weakly stationary.*

PROPOSITION 3.1. *If a Gaussian process is weakly stationary, then it is stationary.*

PROOF. Assume that a Gaussian process X satisfies the above conditions (3.1) and (3.2). Take three integers n, h and k (> 0) arbitrarily, and compare the two distributions of random vectors in Definition 3.1. The two k-dimensional random variables have the same mean vector $(m, m, \ldots, m)$. As for the covariance matrices, their (i, j)-entries are

$$E\{(X_{n+i-1} - m)(X_{n+j-1} - m)\} \quad \text{and} \quad E\{(X_{n+h+i-1} - m)(X_{n+h+j-1} - m)\},$$

respectively, and they are equal to γ_{i-j}. Hence these two random vectors have the same Gaussian distribution. Now use Theorem 2.1 to conclude that X is stationary. □

We now describe some general properties of a weakly stationary process. Let $X = \{X_n(\omega);\ n \in I\}$ be weakly stationary. We denote by $H_n(X)$

the closed linear subspace of $L^2(\Omega, P)$ spanned by the $\{X_k; k \le n\}$. By definition, we obviously have

$$H_n(X) \subset H_{n+1}(X) \quad \text{for every } n \in I. \tag{3.4}$$

If in particular,

$$\bigcap_{n \in I} H_n(X) \subset \{1\} \tag{3.5}$$

holds, then X is said to be *purely nondeterministic* or *regular*. In the above expression $\{1\}$ denotes the subspace of $L^2(\Omega, P)$ which involves only constants, namely $\{1\} = R$.

In contrast to the case above, there is the case where the subspace $H_n(X)$ remains unchanged as n varies, that is

$$H_n(X) = H(X) \quad \text{for every } n \in I.$$

In this case, X is said to be *deterministic*. A weakly stationary process is called *nondeterministic* if it is not deterministic. A purely nondeterministic process is a particular process which is nondeterministic.

The purely nondeterministic case is, of course, more attractive for us from the probabilistic viewpoint, where the relation (3.4) is a proper inclusion. This relation may be replaced by a weaker condition

$$H_n(X) \subsetneqq H_{n+1}(X) \quad \text{holds for some } n, \tag{3.6}$$

since this relation easily implies (3.4) from the assumption that X is stationary.

REMARK 1. We have so far discussed deterministic properties of weakly stationary processes. So far as ordinary (not necessarily Gaussian) stationary processes are concerned, such properties may be discussed in terms of the Hilbert spaces $K_n(X)$ spanned by all the *nonlinear* functions of the X_k, $k \le n$, with finite variance.

REMARK 2. There is another way of defining the deterministic properties of stochastic processes. P. Lévy introduced (in 1953) the notion of a "mixed type" process, which is neither deterministic nor purely nondeterministic. It should be noted that his definition does not use Hilbert spaces such as $H_n(X)$ or $K_n(X)$ but the sample function properties of the given process.

We now return to general weakly stationary processes. To avoid unimportant complexity, we assume in this section the following condition:

$$E(X_n) = 0 \quad \text{for every } n \in I. \tag{A.1}$$

THEOREM 3.1 (H. Wold). *Any weakly stationary process $X = \{X_n(\omega); n \in I\}$ is decomposed into two stationary processes in such a way that*

$$X_n = X_n' + X_n'', \qquad n \in I, \tag{3.7}$$

where $\{X_n'; n \in I\}$ is purely nondeterministic, while $\{X_n''; n \in I\}$ is deterministic. In addition, such a decomposition is unique.

PROOF. First set

$$\bigcap_{n\in I} X_n = H_{-\infty}(X).$$

The projection of X_n down to $H_{-\infty}(X)$ is denoted by X''_n. Set $X'_n = X_n - X''_n$. Thus, the obtained $\{X'_n\}$ and $\{X''_n\}$ are exactly what we wanted to form. The reason for this will be shown below. Let $H_\infty(X)$ be the linear space spanned by all the $\{X_n,\ n \in I\}$, and let U be the unitary operator acting on the space $H_\infty(X)$ determined by

$$UX_n = X_{n+1}, \qquad n \in I. \tag{3.8}$$

Denote by P_k the projection operator from $H_\infty(X)$ down to $H_k(X)$. Then, by definition, we see that $UP_k = P_{k+1}U$, and further it can be proved that

$$U^h P_k = P_{k+h} U^h \tag{3.9}$$

holds for any $h \in I$, where U^{-h} denotes $(U^{-1})^h$. Note that $P_k X_n$ converges to X''_n strongly $k \to \infty$. Hence, we have the equalities

$$U^h X''_n = \lim_{k\to-\infty} U^h P_k X_n = \lim_{k\to-\infty} P_{k+h} X_{n+h} = X''_{n+h},$$

which prove that $X'' = \{X''_n\}$ is weakly stationary.

The above discussion also implies the equality $U^h X'_n = X'_{n+h}$ which proves that $X' = \{X'_n\}$ is weakly stationary and that each X'_n is orthogonal to $H_{-\infty}(X)$. The last assertion and the fact that each X''_n is a member of $H_{-\infty}(X)$ imply that

$$H_n(X'') = H_{-\infty}(X) \quad \text{for every } n \in I, \qquad \text{and} \qquad \bigcap_{n\in I} H_n(X') = \{0\}. \tag{3.10}$$

Hence, H' is purely nondeterministic, while X'' is deterministic.

The uniqueness is almost obvious. In fact, if X' and X'' are purely nondeterministic and deterministic, respectively, then (3.10) shows that X''_n has to be the projection of X_n to $H_{-\infty}(X)$, which completes the proof. □

REMARK. In the above theorem as well as in the proof, we have assumed throughout that all the weakly stationary processes in question are centered, i.e., (A.1).

With the same notation as above, we now have

COROLLARY. *A stationary Gaussian process $X = \{X_n;\ n \in I\}$ admits a decomposition of the form* (3.7), *where X' is purely nondeterministic and X'' is deterministic. The two processes X' and X'' are independent.*

PROOF. First, we show that both X' and X'' are Gaussian. Since X is a Gaussian system, Theorem 2.2 asserts that $H_\infty(X)$ is also Gaussian. Hence, its subsystems X' and X'' are also Gaussian. By Proposition 3.1, we see that they are stationary. In the Gaussian case, orthogonality implies independence so that X' and X'' are independent. □

Given a weakly stationary process, we can subtract off the deterministic part given in the above theorem in order to obtain the interesting component which is the purely nondeterministic part. We may therefore assume that the given process is purely nondeterministic and carry out our analysis.

The notion of canonical representation introduced in the last chapter can easily be generalized to the case where the parameter set is the whole integers I. However we must give additional consideration to the existence and the construction of the representation.

THEOREM 3.2. *A purely nondeterministic stationary Gaussian process* $X = \{X_n(\omega);\ n \in I\}$ *satisfying the assumption* (A.1) *has a canonical representation and the canonical kernel* $a_{n,j}$ *depends only on* $n-j$:

$$a_{n,j} = a_{n-j}, \qquad j \le n. \tag{3.11}$$

PROOF. First, we form the innovation process. It is easy to see that the space $H_n(X)-H_{n-1}(X)$, which is the orthogonal complement of $H_{n-1}(X)$ in $H_n(X)$, is at most one-dimensional. If it is 0-dimensional, we can conclude that $X_n = 0$ for every n, since X is stationary. Hence, $H_n(X) - H_{n-1}(X)$ is exactly one-dimensional, and we may assume that it is spanned by a unit vector ξ_n. Since X is Gaussian, ξ_n is a standard Gaussian random variable. In addition, the system $\Xi = \{\xi_n;\ n \in I\}$, obtained in this manner, forms a Gaussian system. We can also see that the space $H_n(X)$ is spanned by the system $\{\xi_k;\ k \le n\}$. Hence Ξ is the innovation process of X.

With this innovation each X_n is expressed in the form

$$X_n = \sum_{j=-\infty}^{n} a_{n,j}\xi_j, \qquad n \in I.$$

From the construction of ξ_n, we see that

$$E(X_n|X_j,\ j \le k) = P_k X_n = \sum_{j=-\infty}^{k} a_{n,j}\xi_j$$

holds for every $k \le n$. We have therefore obtained a formula which is similar to (2.15) for a process with parameter set I^+.

Since X is stationary, it can be proved that the following two quantities are independent of n.

(i) The variance $a_{n,n}^2$ of $X_n - E(X_n|X_k,\ k \le n-1)$, which does not vanish except for a trivial X, and

(ii) The covariance $\sum_{j=n+1}^{h'} a_{n+h,j}a_{n+h',j}$ of $X_{n+h} - E(X_{n+h}|X_k,\ k \le n)$ and $X_{n+h'} - E(X_{n+h'}|X_k,\ k \le n)$ with $0 < h' < h$.

From (i) the choice of $a_{n,n}$ is unique up to its sign. Choose one of them and denote it by a_0. For (ii) we set $h = 2$ and $h' = 1$. Then $a_0 \cdot a_{n+2,n+1}$ is independent of n, so that $a_{n+2,n+1}$ must be a constant independent of n

and is uniquely determined including its sign. Denote it by a_1. By choosing h and h' successively, we we can prove (3.11). As is seen from such a construction of the sequence $\{a_n\}$, the canonical kernel a_{n-j} is uniquely determined up to a simultaneous change of the signs of a_n. □

EXAMPLE. If, in Example 2 in §2.4, we change the parameter set I^+ to the entire set I, then we are given a stationary Gaussian process expressed in the form

$$X_n = \sum_{j=-\infty}^{n} ac^{n-j}\xi_j, \qquad n \in I, \ 0 < |c| < 1. \tag{3.12}$$

The representation $\{a \cdot c^{n-j}, \xi_j\}$ is canonical.

The Markov property introduced in the last chapter can also be defined in the case where the parameter set is I, and Theorem 2.6 is true without any essential change.

COROLLARY. *If a stationary Gaussian process $X = \{X_n;\ n \in I\}$ is purely nondeterministic and Markov, then X is a process as is given in the above example.*

PROOF. We may assume that X is nontrivial. By Theorem 3.2, the canonical representation exists and the kernel is expressed in the form (3.11). The Markov property implies that the kernel $a_{n,j}$ is of the form (2.20). Hence, we have $a_{n-j} = a_n b_j = a_{n+1} b_{j+1} = \cdots$, so that we obtain

$$\frac{a_{n+1}}{a_n} = \frac{b_j}{b_{j+1}} = c, \qquad c \text{ being a constant independent of } n, j,$$

where we should note that $a_n \neq 0$ and $b_j \neq 0$. We therefore have

$$a_n = a_0 c^n, \quad b_j = b_0 c^{-j}, \qquad n, j \in I.$$

These imply (3.12). The fact that $|c| < 1$ comes from the assumption that the variance of X_n, which is equal to $\sum_{j \geq 0} a^2 c^{2j}$, is finite. □

§2. Spectral representation of a stationary Gaussian process

A stationary Gaussian process has, in addition to the canonical representation, another one called the spectral representation. Such a representation can be given to more general processes, that is to any weakly stationary process. In order to discuss the spectral representation it is convenient for us to consider representation within the class of complex valued random variables so that Fourier analysis can be applied. In what follows all the Gaussian processes are assumed to be *complex* valued. Most results obtained so far for real Gaussian systems also apply work in the complex case without much modification; however, there is one thing worth mentioning. The covariance given by (3.2) and (3.3) should be replaced by

$$E\{(X_{n+h} - m)\overline{(X_n - m)}\} = \gamma_h, \tag{3.13}$$

(the overbar means "complex conjugate") and

$$\gamma_h = \overline{\gamma}_{-h}, \qquad h \in I, \tag{3.14}$$

respectively. Note that a complex Gaussian process forms a complex Gaussian system so that it can be investigated by using the particular properties of a complex Gaussian system established in §2.3.

The meaning of the spectral representation is illustrated by the following example.

EXAMPLE. Let $X = \{X_n(\omega);\ n \in I\}$ be a complex valued process given by

$$X_n(\omega) = \sum_{k=1}^{N} c_k e^{in\lambda_k} Z_k(\omega), \qquad c_k \in C,\ \lambda_k \in R \text{ constants}, \tag{3.15}$$

where $Z_k(\omega)$, $1 \le k \le N$, are mutually independent complex Gaussian random variables with common distribution and with $E(Z_k) = 0$. Obviously, X is a complex Gaussian process. Let the time variable n be shifted by h. Then, each term of the right side of (3.15) can be written respectively in the form

$$c_k e^{in\lambda_k}\{e^{ih\lambda_k} Z_k(\omega)\}, \qquad 1 \le k \le N. \tag{3.16}$$

While it is easy to see that the system $\{e^{ih\lambda_k} Z_k(\omega);\ 1 \le k \le N\}$ has the same distribution as that of $\{Z_k(\omega);\ 1 \le k \le N\}$. Hence, it is proved that X is stationary. Obviously, $E(X_n) = 0$. As for the covariance function γ_h, we set $E|Z_k|^2 = \sigma^2$ to obtain

$$\gamma_h = E(X_{n+h}\overline{X}_n) = \sigma^2 \sum_{k=1}^{N} |c_k|^2 e^{ih\lambda_k},$$

which can be expressed in the form of a Stieltjes integral:

$$\gamma_h = \int_{-\pi}^{\pi} e^{ih\lambda}\, dF(\lambda), \qquad F(\lambda) = \sigma^2 \sum_{\lambda_k < \lambda} |c_k|^2. \tag{3.17}$$

Let us now observe each term of (3.15) again. From the expression we can see that each term $c_k e^{in\lambda_k} Z_k(\omega)$ can be regarded as an *elementary* stationary process. It can be viewed as a simple harmonic oscillator with fluctuating amplitude, where $2\pi/\lambda_k$ looks like the period (although the time is discrete) and $|c_k|$ plays the role of the weight when oscillators are summed up. The graph of $F(\lambda)$ in Figure 3.1 (next page) illustrates how to sum up oscillators with random amplitude to define X_n. In particular, a complex Gaussian process that is represented as in (3.15) can be characterised completely by $F(\lambda)$. The idea is to express a given complex Gaussian process in the form (3.15) or in a more general form. The expression (3.17) for the covariance function γ_h is in this line.

REMARK 1. If, in the formula (3.15), $Z_k(\omega)$ is replaced by $e^{i\theta_k(\omega)}$, with θ_k uniformly distributed on $[-\pi, \pi]$, then we still have a stationary process

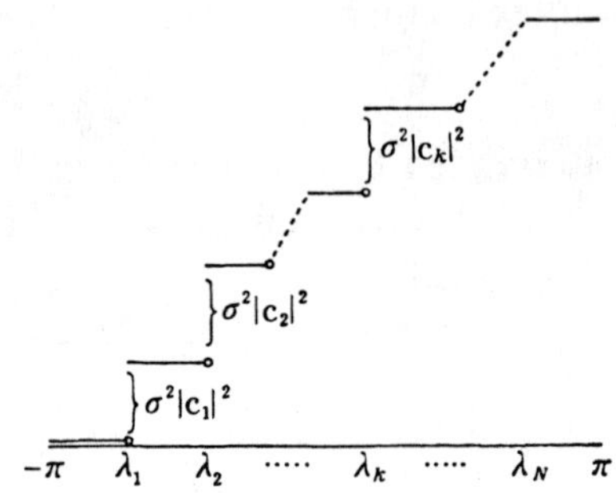

FIGURE 3.1

although it is no longer Gaussian. Such a process is called a *random phase model* and it serves as an important example in applications such as electrical engineering.

REMARK 2. Let all the c_k in the above example be real. Then the process is a complex form (see Definition 2.3.) of a real Gaussian process with mean 0 and with covariance function

$$(\sigma^2/2)\sum_{k=1}^{N} c_k^2 \cos h\lambda_k .$$

REMARK 3. In the above example, if we assume that all the λ_k, $1 \le k \le N$, are different, then with a suitable choice of the n_k, all the Z_k can be expressed as linear combinations of the X_{n_k}. This shows that the process is deterministic.

We are now in a position to discuss the general case. Let $X = \{X_n(\omega);\ n \in I\}$ be a given stationary complex Gaussian process and assume $E(X_n) = 0$ as before. Its covariance function γ_h is positive definite, so we can appeal to the S. Bochner theorem to obtain a spectral representation of the form

$$\gamma_h = \int_{-\pi}^{\pi} e^{ih\lambda}\, dF(\lambda). \tag{3.18}$$

The function $F(\lambda)$ is monotone nondecreasing (hence, dF is a measure) and right continuous with $F(-\pi) = 0$. Such a function $F(\lambda)$ is uniquely determined for the given covariance function γ_h. Naturally, $F(\pi) = \gamma_0$ must hold.

DEFINITION 3.3. The integral in the formula (3.18) is called the *spectral representation* of γ_h, and $F(\lambda)$ is called the *spectral distribution function* of X. If the associated Stieltjes measure $dF(\lambda)$ is absolutely continuous with respect to the Lebesgue measure, then $f(\lambda) = dF(\lambda)/d\lambda$ is the *spectral density function.*

As is well known in measure theory, a spectral distriubtion function F can be decomposed into three parts:

$$F(\lambda) = F_a(\lambda) + F_s(\lambda) + F_d(\lambda), \tag{3.19}$$

where F_a, F_s, and F_d are absolutely continuous, singular (continuous) and discrete parts of F, respectively. Such a decomposition is unique.

The spectral distribution function F in (3.17) for a particular stationary process X given by (3.15) is a good example of the general spectral distribution. Indeed, the F in (3.17) involves only the discrete component. With this observation we may think of a decomposition of the given process itself as a generalization of the expression (3.15). The X_n will now take complex values and so the spaces $H_n(X)$, defined in the same manner as before, form a sequence

$$H_{-\infty}(X) \subset \cdots \subset H_n(X) \subset H_{n+1}(X) \subset \cdots \subset H_\infty(X).$$

A unitary operator U on $H_\infty(X)$ such that

$$UX_n = X_{n+1}, \qquad n \in I$$

is also introduced. Denote its spectral representation by

$$U = \int_{-\pi}^{\pi} e^{i\lambda}\, dE(\lambda), \tag{3.20}$$

where $\{E(\lambda);\ \lambda \in [-\pi, \pi]\}$ is a resolution of the identity on $H_\infty(X)$ satisfying the condition that each $E(\lambda)$ is a projection operator and

$$\begin{aligned} &E(-\pi) = 0, \quad E(\pi) = I \text{ the identity, and} \\ &E(\lambda)E(\mu) = E(\mu)E(\lambda) = E(\lambda \wedge \mu), \quad \lambda \wedge \mu = \min\{\lambda, \mu\}. \end{aligned} \tag{3.21}$$

With these $E(\lambda)$ we introduce a family $Z = \{Z(\lambda, \omega);\ \lambda \in [-\pi, \pi]\}$ of random variables by $Z(\lambda) = E(\lambda)X_0$. We have so far applied only linear operations on the X_n, so that Z forms a complex Gaussian system. For an interval $\Delta = (\lambda, \mu]$ we set $\Delta E = E(\mu) - E(\lambda)$ and $\Delta Z = Z(\mu) - Z(\lambda)$. If Δ_1 and Δ_2 are disjoint, then by the second part of (3.21), $\Delta_1 E$ and $\Delta_2 E$ are orthogonal, and hence $\Delta_1 Z$ and $\Delta_2 Z$ are orthogonal in $H_\infty(X)$. Since Z is a complex Gaussian system, these two increments are independent.

We are now given a complex Gaussian random measure $\{dZ(\lambda)\}$, with which a finite Borel measure $m(d\lambda) = E\{|dZ(\lambda)|^2\}$ is associated (cf. Example 2 in §2.3). A stochastic integral

$$\int_{-\pi}^{\pi} f(\lambda)\, dZ(\lambda) \qquad \text{for } f \in L^2([-\pi, \pi], m)$$

can be defined (see §2.5). Note that $X_n = UX_0^n$ and $U^n = \int_{-\pi}^{\pi} e^{in\lambda}\, dE(\lambda)$. We obtain

$$X_n = \int_{-\pi}^{\pi} e^{in\lambda}\, dZ(\lambda), \qquad n \in I. \tag{3.22}$$

This is the formula we expected as a generalization of (3.15).

DEFINITION 3.4. The representation of $X = \{X_n(\omega);\ n \in I\}$ of the form (3.22) is called the *spectral representation* of X and each X_n is said to have a *spectral decomposition*.

Once the method for obtaining the spectral decomposition of X is shown, the relationship between (3.18) and (3.22) is obvious. In fact, we have

$$\gamma_h = E(X_{n+h}\overline{X_n}) = E\left\{\int_{-\pi}^{\pi} e^{i(n+h)\lambda}\, dZ(\lambda) \int_{-\pi}^{\pi} e^{-in\mu}\, d\overline{Z(\mu)}\right\}$$
$$= \int_{-\pi}^{\pi} e^{ih\lambda} E\{|dZ(\lambda)|^2\} = \int_{-\pi}^{\pi} e^{ih\lambda}\, m(d\lambda),$$

so we can conclude that

$$m(d\lambda) = dF(\lambda),$$

by using the uniqueness property of the spectral measure associated with the covariance function γ_h.

§3. Canonical representation of stationary processes I: The discrete parameter case

The purpose of this section is to construct the canonical representation of a stationary Gaussian process with the help of spectral representation theory. We assume that a given stationary complex Gaussian process $X = \{X_n(\omega);\ n \in I\}$ satisfies

$$E(X_n) = 0, \qquad n \in I. \tag{A.1}$$

Let the spectral representation be given by

$$X_n = \int_{-\pi}^{\pi} e^{in\lambda}\, dZ(\lambda), \qquad E\{|dZ(\lambda)|^2\} = dF(\lambda) \tag{3.23}$$

and introduce the Hilbert space $L^2(F)$ of functions on $[-\pi, \pi]$ given by

$$L^2(F) = \left\{f;\ \int_{-\pi}^{\pi} |f(\lambda)|^2\, dF(\lambda) < \infty\right\}.$$

Then the correspondence

$$X_n \longleftrightarrow e^{i\lambda n} \in L^2(F) \tag{3.24}$$

can be extended linearly to an isomorphism between $H_\infty(X)$ (which is a complex Guassian system) and $L^2(F)$. With a function $h(\lambda)$ and $L^2(F)$ we can associate a stochastic integral

$$\int_{-\pi}^{\pi} h(\lambda)\, dZ(\lambda)$$

in $H_\infty(X)$. The unitary operator U given by (3.8) acts in such a way that

$$U^n \int_{-\pi}^{\pi} h(\lambda)\, dZ(\lambda) = \int_{-\pi}^{\pi} e^{in\lambda} h(\lambda)\, dZ(\lambda), \qquad n \in I.$$

Thus, what we have discussed about $H_\infty(X)$ can be transferred to $L^2(F)$. This means that we deal with ordinary (nonrandom) functions.

With this background we now proceed to a construction of the canonical representation. To guarantee existence we have to assume

$$X \text{ is purely nondeterministic.} \tag{A.2}$$

(See Theorem 3.2. (A.2) is in fact a necessary condition for the existence of a canonical representation.) Under this assumption we can form the innovation process $\{\xi_n;\ n \in I\}$, with which each X_n is to be expressed in the form

$$X_n = \sum_{j=-\infty}^{n} c_{n-j}\xi_j, \qquad \sum_{j=0}^{\infty} |c_j|^2 < \infty. \tag{3.25}$$

On the other hand, since $\{\xi_n\}$ is an independent system, its spectral measure $dF(\lambda)$ in (3.23) has to be of the form $(1/2\pi)\cdot d\lambda$ and there $e^{in\lambda}$ corresponds to ξ_n for every n. Hence, with the right-hand side of (3.25) we can associate an $L^2([-\pi, \pi])$-element

$$\sum_{j=-\infty}^{n} c_{n-j}e^{ij\lambda} = e^{in\lambda}\sum_{k=0}^{\infty} c_k e^{-ik\lambda}.$$

Denote the right-hand side by $e^{in\lambda}\varphi(e^{-i\lambda})$. Using this notation we find the corresponding function to y_h in the space $L^2(F)$ and compare with (3.18) to obtain the expression

$$dF(\lambda) = (2\pi)^{-1}|\varphi(e^{-i\lambda})|^2\, d\lambda. \tag{3.26}$$

In particular, it has been proved that $dF(\lambda)$ is absolutely continuous under the assumption (A.2). In addition, we should note that $\varphi(e^{i\lambda})$ is in the subspace of $L^2([-\pi, \pi])$ spanned by the system $\{e^{ik\lambda};\ k \in I^+\}$. In fact, $\{c_k;\ k \geq 0\}$ is the set of Fourier coefficients of the function $\varphi(e^{i\lambda})$.

From what we have discussed so far it follows that in order to obtain the kernel $\{c_{n-j}\}$ of the representation (3.25) it suffices for us to find a square root φ of the spectral density function. Note that there is much freedom left in choosing a square root within the complex numbers. In fact, in constructing the canonical representation we are choosing a suitable square root among many candidates. The construction is shown below in a step by step manner.

(i) We observe necessary conditions for φ to be the square root that gives the canonical kernel. First, $\varphi(0) \neq 0$ must hold. For, $\varphi(0) = 0$ implies that X_n is expressed as a function of the ξ_k, $k \leq n-1$, which leads us to a contradiction. Now set $\varphi(0) = c_0$ $(\neq 0)$. Then our assumption that (3.25) gives the canonical representation implies

$$E[\{X_n - E(X_n|B_{n-1}(X))\}^2] = |c_0|^2.$$

Hence, the modulus of c_0 is determined. Let the argument of c_0 be undecided. Next it is seen that $\varphi(z)$ never vanishes within the unit disc $|z| < 1$.

In fact, suppose that $\varphi(z_0) = 0$ were true for some z_0 with $0 < |z_0| < 1$. Then setting $X' = \sum_{j\le n} z_0^j \xi_{n-j}$, we see that $X' \in H_n(\xi)$ $(= H_n(X))$, but X' is not in $H_{n-1}(X)$ $(z_0 \neq 0)$. On the other hand, we have

$$E(X' \cdot \overline{X}_n) = \sum_{j=0}^{\infty} \overline{z}_0^j E(\xi_{n-1} \cdot \overline{X}_n) = \sum_{j=0}^{\infty} \overline{z}_0^j \overline{c}_j = \overline{\varphi(z_0)} = 0.$$

Hence, X_n is orthogonal to X', that is independent of X'. This is a contradiction.

We have therefore proved that $\varphi(z)$ is analytic on the unit disc and never vanishes there.

(ii) We next prove that $\log|\varphi(e^{i\lambda})|$ is integrable. Since $\log\varphi(z)$ is analytic (by (i)) on the domain $|z| < 1$, we have

$$\log\varphi(0) = \frac{1}{2\pi i}\int_{|z|=r} \frac{\log\varphi(z)}{z}\,dz, \qquad 0 < r < 1.$$

Taking the real parts of both sides, we obtain

$$\log|\varphi(0)| = \frac{1}{2\pi}\int_{-\pi}^{\pi} \log|\varphi(re^{i\lambda})|\,d\lambda.$$

From the fact that $|\varphi(e^{i\lambda})|$ is integrable (this fact comes from the relation (3.26)) and from Jensen's inequality, the following inequalities are proved:

$$\int_{-\pi}^{\pi} \log|\varphi(e^{i\lambda})|\,d\lambda \le \log\int_{-\pi}^{\pi} |\varphi(e^{i\lambda})|\,d\lambda < \infty.$$

Now the above expression for $\log|\varphi(0)|$ gives us

$$-\frac{1}{2\pi}\int_{-\pi}^{\pi} \log^{-}|\varphi(re^{i\lambda})|\,d\lambda \le -\log|\varphi(0)| + \frac{1}{2\pi}\int_{-\pi}^{\pi} \log^{+}|\varphi(re^{i\lambda})|\,d\lambda < \infty.$$

(For the notations: $\log^{-} x = \operatorname{Min}\{\log x, 0\}$ and $\log^{+} x = \operatorname{Max}\{\log x, 0\}$.) Take the $\limsup_{r\to 1-}$ of the right side and $\liminf_{r\to 1-}$ of the left side. Then, we have

$$\liminf_{r\to 1-}\left\{-\frac{1}{2\pi}\int_{-\pi}^{\pi} \log^{-}|\varphi(re^{i\lambda})|\,d\lambda\right\}$$
$$\le \left\{-\log|\varphi(0)| + \limsup_{r\to 1-}\frac{1}{2\pi}\int_{-\pi}^{\pi} \log^{+}|\varphi(re^{i\lambda})|d\lambda\right\}.$$

The Jensen inequality guarantees

$$-\frac{1}{2\pi}\int_{-\pi}^{\pi} \log^{-}|\varphi(e^{i\lambda})|\,d\lambda \le -\log|\varphi(0)| + \frac{1}{2\pi}\int_{-\pi}^{\pi} \log^{+}|\varphi(e^{i\lambda})|\,d\lambda < \infty.$$

Note that the last inequality comes from the integrability of the function $\log|\varphi(e^{i\lambda})|$.

Together with the preceeding evaluation, we have now proved

$$-\infty < \int_{-\pi}^{\pi} \log|\varphi(e^{i\lambda})|d\lambda < \infty.$$

(iii) Let $f(\lambda)$ denote the spectral density $dF(\lambda)/d\lambda$. From the earlier discussion it must equal $|\varphi(e^{i\lambda})|^2/2\pi$, so that the function $\log f(\lambda)/2$ is integrable. It should be kept in mind that $f(\lambda)$ is a function obtained from the spectral representation. Now let

$$\frac{1}{2}\log f(\lambda) \sim \sum_{n=-\infty}^{\infty} a_n e^{in\lambda}$$

be the Fourier series expansion, where

$$a_n = \frac{1}{4\pi}\int_{-\pi}^{\pi} \log f(\lambda) e^{-in\lambda}\, d\lambda \quad \text{and} \quad a_{-n} = \overline{a_n}.$$

With these a_n we define a function $g(z)$ by

$$g(z) = a_0 + 2\sum_{n-1}^{\infty} a_n z^n, \qquad |z| < 1.$$

Then $g(z)$ is analytic on the unit disc and

$$\operatorname{Re} g(re^{i\lambda}) = \sum_{n=-\infty}^{\infty} a_n r^{|n|} e^{in\lambda} \quad \text{(Re means the real part)}$$

is a harmonic function for $r < 1$ with boundary value $\log f(\lambda)/2$. We can determine a sequence $\{c'_j;\ j \geq 0\}$ by

$$(2\pi)^{1/2}\exp[g(z)] = \sum_{j=0}^{\infty} \overline{c'_j z^j}$$

and define an analytic function

$$c(z) = \sum_{j=0}^{\infty} c'_j z^j, \qquad |z| < 1. \tag{3.27}$$

The function $c(z)$ defined above belongs to the Hardy class $\mathbf{H}_2$, and $c(z)$ has no zero inside of the unit disc. These facts are proved as follows.

$$\int_{-\pi}^{\pi} |c(re^{-i\lambda})|^2\, d\lambda = 2\pi \int_{-\pi}^{\pi} \exp[2\operatorname{Re} g(re^{i\theta})]\, d\theta \qquad (0 < r < 1)$$

$$= 2\pi \int_{-\pi}^{\pi} \exp\left[\frac{1}{2\pi}\int_{-\pi}^{\pi} \frac{1-r^2}{1-2r\cos(\theta-\lambda)+r^2} \log f(\lambda)\, d\lambda\right] d\theta$$

(by the Poisson formula)

$$\leq 2\pi \int_{-\pi}^{\pi} f(\lambda) \left\{\frac{1}{2\pi}\int_{-\pi}^{\pi} \frac{1-r^2}{1-2r\cos(\theta-\lambda)+r^2}\, d\theta\right\} d\lambda$$

$$= 2\pi \int_{-\pi}^{\pi} f(\lambda)\, d\lambda.$$

Hence, the boundary value of $c(x)$ exists and is given by

$$\lim_{r\to 1-} c(re^{-i\lambda}) = c(e^{-i\lambda}) = \sum_{j=0}^{\infty} c'_j e^{-ij\lambda},$$

with

$$\frac{1}{2\pi}|c(e^{-i\lambda})|^2 = f(\lambda).$$

(iv) As for the function $\varphi(z)$, we note that $\log|\varphi(z)|$ is a harmonic function on the unit disc in order to have

$$\log|\varphi(re^{i\lambda})| = \frac{1}{2\pi}\int_{-\pi}^{\pi} \frac{R^2 - r^2}{R^2 - 2Rr\cos(\theta - \lambda) + r^2} \log|\varphi(Re^{i\theta})|\, d\theta \qquad (0 < r < R < 1)$$

$$\le \frac{1}{2\pi}\int_{-\pi}^{\pi} \frac{1 - r^2}{1 - 2r\cos(\theta - \lambda) + r^2} \log|\varphi(e^{i\theta})|\, d\theta \qquad \text{(Fatou's lemma)}$$

$$= \frac{1}{2}\log 2\pi + \frac{1}{4\pi}\int_{-\pi}^{\pi} \frac{1 - r^2}{1 - 2r\cos(\theta - \lambda) + r^2} \log f(-\theta)\, d\theta$$

$$= \log|c(re^{i\lambda})|.$$

We have therefore proved that $\log|c(re^{i\lambda})/\varphi(re^{i\lambda})|$ is nonnegative and harmonic.

(v) We now return to the one-to-one correspondence (3.24). Choose an element

$$e^{i\lambda}\sum_{j=1}^{\infty} c_j' e^{-ij\lambda}/c(e^{-i\lambda}) \tag{3.28}$$

in the space $L^2(F)$. It belongs to the subspace $L_0(F)$ of $L^2(F)$ spanned by $\{e^{ij\lambda};\ j \le 0\}$, since $c(z) \ne 0$, for $|z| < 1$. On the other hand, it is easy to see that the function

$$e^{i\lambda}\left(1 - \sum_{j=1}^{\infty} c_j' e^{-ij\lambda}\right) \Big/ c(e^{-i\lambda}) \tag{3.29}$$

is orthogonal to $L_0(F)$. Because the space $L_0(F)$ is spanned by $\{e^{ik\lambda};\ k \le 0\}$ and the inner product below vanishes:

$$\int_{-\pi}^{\pi} e^{i\lambda}\left\{1 - \sum_{j=1}^{\infty} c_j' e^{-ij\lambda}/c(e^{-i\lambda})\right\} \cdot \overline{\{e^{ik\lambda}/c(e^{-i\lambda})\}} f(\lambda)\, d\lambda$$

$$= \int_{-\pi}^{\pi} e^{i(1-k)\lambda}\left\{c(e^{-i\lambda}) - \sum_{j=1}^{\infty} c_j' e^{-ij\lambda}\right\} d\lambda = 0, \qquad k \le 0.$$

These results can be rephrased in terms of $H(X)$; the element (3.28) turns into the projection of X_1 to the subspace $H_0(X)$ and that of (3.29) is accordingly the component of the X_1 orthogonal to $H_0(X)$. The variance of the latter can be computed to have

$$\int_{-\pi}^{\pi} \left|1 - \sum_{j=1}^{\infty} c_j' e^{-ij\lambda}/c(e^{-i\lambda})\right|^2 f(\lambda)\, d\lambda = |c_0'|^2 = |c(0)|^2. \tag{3.30}$$

(We shall return to this formula later.) This implies that $\log|c(0)/\varphi(0)| = 0$. Referring to the result in (iv), it is proved that $h(z) = \log|c(z)/\varphi(z)|$ is identically 0 on the unit disc, since $h(z)$ is a nonnegative harmonic function which vanishes at 0. Namely, $\log|c(z)| = \log|\varphi(z)|$ for $|z| < 1$. Thus, the two analytic functions can differ only by a constant factor $e^{i\alpha}$ with α real

$$c_j' = e^{i\alpha} c_j, \qquad j \geq 0.$$

Such a difference is a part of an equivalence.

(vi) The final step is devoted to a construction of the innovation, although the essential part of the construction has been done already in step (v). By using the correspondence (3.24), the ξ_n that we are looking for is determined as the element corresponding to the function $e^{in\lambda}/c(e^{-i\lambda})$. Hence, if we write

$$\frac{1}{c(e^{-i\lambda})} = \sum_{k=0}^{\infty} d_k e^{-ik\lambda},$$

then ξ_n is given by

$$\xi_n = \sum_{k=0}^{\infty} d_k X_{n-k}. \tag{3.31}$$

Summing up what we have discussed, we have proved the following theorem.

THEOREM 3.3. *Let $X = \{X_n(\omega)\,;\, n \in I\}$ be a complex stationary Gaussian process satisfying the assumptions* (A.1) *and* (A.2). *Then*

(i) *The spectral measure is absolutely continuous and the density function $f(\lambda)$ satisfies the condition*

$$\left| \int_{-\pi}^{\pi} \log f(\lambda)\, d\lambda \right| < \infty.$$

(ii) *The canonical representation exists and the canonical kernel $\{c_{n-j},\, j \leq n\}$ is uniquely determined up to a common factor with modulus one. The c_j appear as coefficients of $c(z)$ as in* (3.27) (*where c_j' is replaced by c_j*).

(iii) *The innovation sequence $\{\xi_n\}$ is formed by* (3.31).

There are two supplementary remarks to Theorem 3.3.

(1) Before coming to the conclusion of the theorem, we followed the steps (i)–(v) in order to form the kernel $\{c_{n-j}\,;\, j \leq n\}$. In the course of these steps we were able to explain our ideas, however once one knows the whole story it is convenient to form $c(e^{-i\lambda})$ directly from the spectral density function $f(\lambda)$. There is a formula for this purpose:

$$c(z) = \sum_{j=0}^{\infty} c_j z^j = \sqrt{2\pi} \exp\left[\frac{1}{4\pi} \int_{-\pi}^{\pi} \frac{e^{-i\theta} + z}{e^{-i\theta} - z} \log f(\theta)\, d\theta \right]. \tag{3.32}$$

(2) There is a characterization of the function $c(z)$ which gives the canonical kernel. Namely, consider the class **A** of all the $a(z)$ in the Hardy class $\mathbf{H}_2$ such that

$$a(0) > 0 \quad \text{and} \quad \frac{1}{2\pi}|a(e^{-i\lambda})|^2 = f(\lambda).$$

Then, our favorite function $c(z)$ is the one in **A** with maximum modulus.

EXAMPLE. Let $X = \{X_n(\omega);\ n \in I\}$ be a complex stationary Gaussian process with spectral density function $f(\lambda)$ given by

$$f(\lambda) = |1 - \alpha e^{-i\lambda}|^{-2}, \qquad |\alpha| < 1. \tag{3.33}$$

Then the function $c(z)$ determined by (3.32) is

$$c(z) = \frac{1}{1 - \alpha z}.$$

Thus, the canonical kernel $\{c_{n-j}\}$ obtained from $c(z)$ has the form

$$c_{n-j} = \alpha^{n-j}.$$

It might be interesting to give an example which is expressed as a moving average, but the expression does not give the canonical representation. The following function $a(z)$ gives us such an example:

$$a(z) = \frac{\beta - z}{(1 - \alpha z)(1 - \beta z)}, \qquad 0 < \beta < 1.$$

In this case, the proper inclusion

$$H_n(X) \subsetneqq H_n(\xi)$$

holds for every $n \in I$.

§4. Canonical representation of stationary processes II: The continuous parameter case

This section is devoted to the canonical representation theory of a stationary complex Gaussian process $X = \{X(t, \omega);\ t \in R\}$ with the continuous parameter t. Since the idea for obtaining the representation is the same as that in the previous section for processes with the discrete parameter n, we describe only the results which emphasize the differences from results in the previous section.

The story starts with the spectral representation of X as in the last section. We use the following notation:

$H_\infty(X) =$ the subspace of $L^2(\Omega, P)$ spanned by $\{X(s);\ s \in R\}$,
$H_t(X) =$ the subspace of $H_\infty(X)$ spanned by $\{X(s);\ s \le t\}$,
$H_{-\infty}(X) = \bigcap_t H_t(X)$.

As in the case of (3.8), a unitary operator U_t on $H_\infty(X)$ such that

$$(U_t X)(s) = X(s + t), \qquad t \in R$$

can be defined and $\{U_t\,;\ t \in R\}$ forms a one-parameter group of unitary operators acting on $H_\infty(X)$,

$$U_t U_s = U_{t+s}, \qquad t, s \in R.$$

In what follows we always assume

$$X(t) \text{ is mean square continuous in } t. \tag{A.3}$$

With this assumption, U_t is strongly continuous in t. Hence, Stone's theorem for the continuous one-parameter unitary group $\{U_t\}$ yields the spectral decomposition of U_t:

$$U_t = \int_{-\infty}^{\infty} e^{it\lambda}\, dE(\lambda),$$

where $\{E(\lambda)\,;\ \lambda \in R\}$ is the resolution of the identity on $H_\infty(X)$.

Now set $E(\lambda)X(0) = Z(\lambda)$. Then $Z = \{Z(\lambda, \omega)\,;\ \lambda \in R\}$ is a complex Gaussian process with independent increments, and $X(t) = U_t X(0)$ itself can be expressed as a Wiener integral of the form

$$X(t) = \int_{-\infty}^{\infty} e^{i\lambda t}\, dZ(\lambda) \tag{3.34}$$

which comes immediately from the spectral decomposition of U_t. Formula (3.34) is called the *spectral representation* of $X(t)$. This is the counterpart of the representation (3.23) for the discrete parameter case.

The spectral measure $dF(\lambda)$ of the stationary process X is given by the formula

$$E(|dZ(\lambda)|^2) = dF(\lambda).$$

By using this measure (in fact, using the density function), we shall form the canonical kernel, $G(t-u)$, $u \le t$, which is to be a function of $t-u$ only.

Remember our assumptions. We assume conditions obtained by modifying (A.1) and (A.2) (see the last section) so that they are conditions for a continuous parameter processes. In addition, condition (A.3) is assumed. $L^2(F)$ is the space spanned by $\{e^{it\lambda}\,;\ t \in R\}$, and the spectral measure $dF(\lambda)$ is absolutely continuous with density $f(\lambda)$. The Hardy space $\mathbf{H}_2$, where the preferred square root of $f(\lambda)$ can be found, involves analytic functions over the lower half plane. With these remarks we can state the main theorem.

THEOREM 3.4. *If a stationary complex Gaussian process* $\{X(t, \omega)\,;\ t \in R\}$ *satisfies assumptions* (A.1), (A.2), *and* (A.3), *then*

(i) *the spectral measure is absolutely-continuous and the density function* $f(\lambda)$ *satisfies the inequality*

$$\int_{-\infty}^{\infty} \frac{\log f(\lambda)}{1+\lambda^2}\, d\lambda > -\infty.$$

(ii) *The canonical kernel is obtained as follows. Take the function*

$$c(w) = \sqrt{2\pi}\exp\left[-\frac{1}{2\pi i}\int_{-\infty}^{\infty} \frac{1+\lambda w}{\lambda - w}\,\frac{\log f(\lambda)}{1+\lambda^2}\, d\lambda\right], \tag{3.35}$$

which is in $\mathbf{H}_2$, *and the limit*

$$c(\lambda) = \lim_{\mu \to 0-} c(\lambda + i\mu).$$

The canonical kernel $G(t-u)$ *is obtained from the Fourier transform of the* $c(\lambda)$ *in such a way that*

$$G(u) = \frac{1}{2\pi} \int_{-\infty}^{\infty} e^{iu\lambda} c(\lambda)\, d\lambda.$$

REMARK 1. It can easily be seen, having been suggested by the results in the last section, that

$$f(\lambda) = \frac{1}{2\pi} |c(\lambda)|^2 \quad \text{and} \quad G(u) = 0 \quad \text{for } u < 0.$$

The explicit form of the canonical representation of $X(t)$ follows immediately from the above theorem. Namely, there exists a complex Gaussian random measure that comes from a complex Brownian motion and, by using the kernel $G(t-u)$ given by the above theorem, $X(t)$ can be expressed in the form

$$X(t) = \int_{-\infty}^{t} G(t-u)\, dB(u). \tag{3.36}$$

REMARK 2. It is difficult to have an analogue of Part (iii) Theorem 3.3 for forming the innovation from the function $c(\lambda)$. Unlike the discrete parameter case, the kernel $G(t-u)$ should be viewed as an integral operator, so that it is not easy to find an explicit form for the inverse operator.

REMARK 3. The counter part of the function $a(z)$ that appeared in Remark 2 following Theorem 3.3 is described in such a way that a possible expression of $a(w)$ in the space $\mathbf{H}_2$ is

$$a(w) = c(w) \prod(w) \cdot \exp\left[\frac{1}{2\pi i} \int_{-\infty}^{\infty} \frac{1 + \lambda w}{\lambda - w}\, d\chi(\lambda) - i\beta w + i\alpha\right], \tag{3.37}$$

where $\prod(w)$ is the Blaschke product, $d\chi(\lambda)$ is a singular measure and β, α are real constants.

There is an important note regarding complex Gaussian system. As in the last section, we have started with a complex Gaussian process. The random variables appearing in our discussion are all in the Hilbert space $H_\infty(X)$. They are still complex Gaussian and their probability distributions can easily be specified. In particular, so far as we remain within complex Gaussian system orthogonality is equivalent to independence. One can ask what to do when given a real Gaussian process. The answer is to first form its complex version which is a complex Gaussian system. After the canonical representation is obtained, one can take the real part, which is exactly what is sought. In this case, the canonical kernel is real valued and the integral is the (real) Wiener integral.

CHAPTER IV

Canonical Representation of Gaussian Processes: General Theory and Multiplicity

§1. Propagation of random phenomena

We start with a study, from the view point of the theory of stochastic processes, of mathematical models of random phenomena that change with time. The models that we discuss in this chapter are all Gaussian processes that depend on a continuous parameter denoted by $X = \{X(t, \omega);\ t \in I\}$, I being an interval.

In the continuous parameter case, the processes in question enjoy a more complex structure, as is seen from their behaviors, compared to the discrete parameter case. Among others, there arises an important new concept, which will be introduced later, the *multiplicity* of a Gaussian process.

There are several powerful probabilisitic methods for investigating how the given randomness propagates and for expressing the characteristics presented when the development of the time is taken into account. We now list some of the well-known methods below.

(1°) Let a Gaussian process $\{X(t)\}$ be given. The most basic method is to arbitrarily choose n points t_i, $1 \le i \le n$, and observe the behavior of the joint distribution $P(X(t_1) \le x_1, X(t_2) \le x_2, \dots, X(t_n) \le x_n)$ when the t_i change. This technique is most useful if $\{X(t)\}$ is a Markov process. In this case, the joint distribution can be obtained from conditional probabilities $P(X(t_{j+1}) \le x_{j+1} | X(t_j))$, $t_{j+1} > t_j$, which are determined only by pairs of the t_j's. Thus, computing the joint distribution is easy, and further it is possible to give an intuitive interpretation in such a way that if the value of the present time is known, then the future and the past values become mutually independent. This observation enables us to come to a generalization of the Markov property, namely the multiple Markov property, which is discussed in detail in Chapter 5.

(2°) Since the time parameter t changes continuously, we will be concerned with analytic properties of random phenomena in terms of the continuity or differentiability in t of the sample function, for a given process. More generally we are interested in regularity of sample functions $X(t, \omega)$ which are functions of t for a fixed ω. Concerning this topic, there are many good references specifically for Gaussian processes, since it is such an important subject. We will, however, not go into the details because the method of investigation is not quite in line with our approach. Nevertheless we shall often meet continuities of $X(t)$ like mean square continuity or continuity in probability throughout our study, and continuity is viewed as one of the properties that describe propagation of random events locally in time.

(3°) The next method, the importance of which should be emphasized most, is the use of the notation of multiplicity for a process as is stated in the title, although this notion is limited to Gaussian processes. There are many ways to give an intuitive interpretation of this notion, while a rigorous definition will be given in the next section in a general setting. A rough idea will be illustrated quickly in what follows.

Now, we recall the notion of the *innovation process* which was introduced in Chapter 2 for discrete parameter Gaussian processes. It contains, at every moment, the same amount of information as the given Gaussian process gains marginally at that moment, and it is independent of the past of the given process. Eventually, the innovation process for a given Gaussian process should, in the discrete parameter case, be a sequence of independent random variables.

One may think of a single white noise, or equivalently, of the system of increments of a Brownian motion which is a continuity analogue of the innovation just reviewed. In reality, such an easy guess is not quite sufficient for our purpose. Since the time variable is continuous, invasion of the innovation may take place with higher multiplicity.

There are various reasons why higher multiplicity appears, however, we know only a few of them. A simple but somewhat artificial example is formed by taking a sequence of disjoint, countable, dense subsets of real numbers, and by associating a mutually independent system of Gaussian random variables with each of those subsets. Then, it is easy to form a continuous parameter Gaussian process in which those systems are imbedded. The process obviously has countably infinite multiplicity as is easily seen. There is another example, which comes from the example in §4, part D. Unlike the above, it is less sophisticated. The construction of the Gaussian process in question, with higher multiplicity, shall use the sample function regularity of Brownian motion. As we have seen in §2.5, sample functions of Brownian motion enjoy a particular regularity uniformly in t; continuous but not differentiable. Now take another Gaussian process with a different sample function regularity so that its sample functions can be discriminated from those of Brownian motion. Add such a Gaussian process and a Brownian motion to form a new Gaussian process; let it be denoted by $Y(t, \omega)$. Then it is easy to see that a sample function $Y(t, \omega)$ for a fixed ω can be decomposed into two functions with different regularity. Those components are of course sample functions of two processes that are summands of $Y(t, \omega)$. In other words, the sum $Y(t, \omega)$ is carrying the information that is contained in two processes without any loss while it is transmitted. Double multiplicity may arise in such a manner. Suppose that this example is a mathematical model of a noise that occurs in an actual phenomenon. Then this suggests to us that a noise can not be dealt with as a *simple* random process, but a mixture, where "simple" means that all the sample functions satisfy a certain analytic property uniformly in ω. Indeed, this example has double multiplicity.

Note that the notion of multiplicity explained in (3°) naturally arises if we wish to discuss a general theory of canonical representation of Gaussian processes. In fact, the multiplicity of a Gaussian process could have never been thought of without a consideration of canonical representation theory.

Having made these preliminary observations on canonical representation and multiplicity we are now ready to begin the main topic.

§2. Canonical representation and multiplicity

Let $X = \{X(t);\ t \in I\}$ be a Gaussian process. Denote by $\mathbf{B}_t(X)$ the σ-field generated by the $X(s)$, $s \le t$. Set $\mathbf{B}_{t+}(X) = \bigcap_{u>t} \mathbf{B}_u(X)$ and $\mathbf{B}_{t-}(X) = \bigvee_{u<t} \mathbf{B}_u(X)$. $\bigvee_{u<t} \mathbf{B}_u(X)$ denotes the smallest σ-field containing all the $\mathbf{B}_u(X)$ with $u < t$. Now we are able to give our definition of the canonical representation.

DEFINITION 4.1. Let X be a Gaussian process. Assume that there exists a Gaussian process $B = \{B(t);\ t \in I\}$ with independent increments such that

$$\begin{aligned} B(t) &= (B_i(t),\ 1 \le i \le N;\ B_j^l(t),\ 1 \le l \le L_j,\ 1 \le j \le J) \\ B(0) &= 0, \qquad N \le \infty,\ L_j \le \infty,\ J \le \infty, \end{aligned} \tag{4.1}$$

for which the following conditions (I.1)–(I-4) are satisfied:

(I.1) Each $B_i(t)$ has independent increments and $E(|dB_i(t)|^2) = m_i(dt)$ defines a continuous measure. In addition, m_{i+1} is absolutely continuous with respect to m_i : $m_i(dt) \gg m_{i+1}(dt)$ for every i.

(I.2) Each $B_j^l(t)$ is a process of the form

$$B_j^l(t) = \begin{cases} 0, & t \le t_j \text{ (or } t < t_j), \\ B_j^l, & t > t_j \text{ (or } t \ge t_j), \end{cases} \tag{4.2}$$

where each B_j^l is subject to the standard Gaussian distribution $N(0, 1)$.

(I.3) The Gaussian processes B_i and the B_j^l are independent.

(I.4) For every t $X(t)$ has the same distribution as $\widetilde{X}(t)$ given by

$$\widetilde{X}(t) = \sum_{i=1}^{N} \int_0^t F_i(t, u)\, dB_i(u) + \sum_{\{t_j \le t\}} \sum_{l=1}^{L_j} b_j^l(t) B_j^l(t), \tag{4.3}$$

where the kernel functions $F_i(t, u)$ satisfy the conditions

$$\sum_{i=1}^{N} \int_0^t F_i(t, u)\, m_i(du) < \infty, \qquad i = 1, 2, \ldots, N,$$

for every t, and where the function $b_j^l(t)$ vanishes for $t_j < t$ and satisfy

$$\sum_{\{t_j \le t\}} \sum_{l=1}^{L_j} b_j^l(t)^2 < \infty \quad \text{for every } t.$$

Then, $\{F_i(t, u), B_i(u); b_j^l(t), B_j^l(t)\}$ is called a *representation* of X. Further, if the equalities $\mathbf{B}_t(\widetilde{X}) = \mathbf{B}_t(B)$, $\mathbf{B}_{t+}(\widetilde{X}) = \mathbf{B}_{t+}(B)$, and $\mathbf{B}_{t-}(\widetilde{X}) = \mathbf{B}_{t-}(B)$ hold for any t, then X is said to be represented *canonically* or X has the *canonical representation* relative to B.

If $X(t)$ itself can be represented canonically in the form (4.3) (i.e. $\widetilde{X}(t) = X(t)$, almost surely), then the process B is called the *innovation process* for X.

Before the concept of multiplicity of a Gaussian process X is introduced it should be noted, with regard to the canonical representation, that the number N in (4.1) is uniquely determined, in fact, it is independent of the choice of the system $\{B_i(t)\}$. This comes from Theorem 4.1 to be proved.

DEFINITION 4.2. The number N in the above representation is called the *continuous multiplicity* of X. L_j is the *discrete multiplicity* at t_j and $L = \sup_{1 \le j \le J} L_j$ is simply called the *discrete multiplicity* of X. The (total) *multiplicity* M of X is defined by $M = \mathrm{Max}\{N, L\}$.

DEFINITION 4.3. The measure $m_i(dt) = E(|dB_i(t)|^2)$ is called a *continuous spectral measure* of X and the system $\{m_i(dt);\ 1 \le i \le N\}$ is called the system of continuous spectral measures.

REMARK. The system of continuous spectral measures is unique up to equivalence. This fact may be seen in the proof of the following theorem.

EXAMPLE. A Brownian motion $B = \{B(t);\ t \in I\}$ is canonically represented by itself. Namely, $B_1(t)$ is taken to be the given $B(t)$ which is $\int_0^t 1\, dB(u)$, $t \in I$. Obviously $N = 1$, $L = 0$, $m_1(dt) = dt$ (The Lebesgue measure). It is interesting to note that there are many noncanonical (namely, the equalities for σ-fields in Definition 4.1 fail to hold) representations of a Brownian motion. We shall see this later.

With the notations established in §3.4, we assume the following three conditions throughout this chapter.

(A.1) $E(X(t)) \equiv 0$,
(A.2) $\bigcap_t H_t(X) = \{0\}$, that is, purely nondeterministic,
(A.3) $H_\infty(X)$ is separable.

THEOREM 4.1. (I) *For a Gaussian process* $X = \{X(t);\ t \in I\}$ *with* $I = [0, \infty)$, *a canonical representation of the form* (4.3) *exists.*

(II) (*Uniqueness*) *If* X *has another canonical representation of the form*

$$\overline{X}(t) = \sum_{i=1}^{\overline{N}} \int_0^t \overline{F}_i(t, u)\, d\overline{B}_i(u) + \sum_{\{s_k \le t\}} \sum_{l=1}^{\overline{L}_k} \overline{b}_k^l(t) \overline{B}_k^l(t), \tag{4.3$'$}$$

then we have that $N = \overline{N}$, $m_i \sim \overline{m}_i$ (*equivalent*) *with* $\overline{m}_i(du) = E(|d\overline{B}_i(u)|^2)$, $i \ge 1$, *the set* $\{t_j\}$ *is the same as* $\{s_k\}$, *and* $L_j = \overline{L}_k$ *if* $t_j = s_k$.

REMARK. In Theorem 4.1. the time interval I is taken to be $[0, \infty)$. There arises no essential difference if I is changed to a finite interval or to $(-\infty, \infty)$, as we shall see later.

PROOF OF THEOREM. (i) Since X is separable, there is a countable set $\{\tau_i \in I;\ i = 1, 2, \dots\}$ such that

$$\sigma(X(\tau_1), X(\tau_2), \dots, X(\tau_n), \dots) = \mathbf{B}(X),$$

where $\mathbf{B}(X)$ is the σ-field generated by $\{X(t);\ t \in I\}$. We set

$$M_i(t) = E(X(\tau_i)|\mathbf{B}_t(X)), \qquad M_i(t\pm) = E(X(\tau_i)|\mathbf{B}_{t\pm}(X)).$$

It is noted that $M_i = \{M_i(t), \mathbf{B}_t(X);\ t \in I^*\}$ is a Gaussian martingale, where I^* is an ordered set defined by

$$I^* = I \cup \{t-;\ t \in I\} \cup \{t+;\ t \in I\}$$

with the order such that for $t < t'$, the inequalities $t- < t < t+ < t'- < t' < t'+$ should hold.

(ii) Define $\mathbf{B}_t(M_i) = \sigma\{M_i(s);\ s \le t\}$. Similarly $\mathbf{B}_{t-}(M_i)$ and $\mathbf{B}_{t+}(M_i)$ are defined. We then form a system $\{N_i;\ i = 1, 2, \dots\}$ of mutually independent Gaussian martingales from the system $\{M_i;\ i = 1, 2, \dots\}$ of martingales given in (i). Now we set

$$N_1(t) = M_1(t), \qquad t \in I^*.$$

For $i \ge 2$, $N_i(t)$ is defined inductively in the following manner:

$$N_i(t) = M_i(t) - E\left\{ M_i(t)| \bigvee_{j \le i-1} \mathbf{B}_t(M_j) \right\}, \qquad t \in I,$$

$$N_i(t\pm) = M(t\pm) - E\left\{ M_i(t\pm)| \bigvee_{j \le i-1} \mathbf{B}_{t\pm}(M_j) \right\}, \qquad t \in I.$$

It is easy to see that the Gaussian martingales $N_j = \{N_j(t), \mathbf{B}_t(X);\ t \in I^*\}$, $j = 1, 2, \dots$ are mutually independent and that

$$\bigvee_{j \le i} \mathbf{B}_t(M_j) = \bigvee_{j \le i} \mathbf{B}_t(N_j), \qquad t \in I^*,$$

holds for every i, where $\mathbf{B}_t(N_j) = \sigma\{N_j(s); s \le t\}$. Here we only show that N_i is a martingale. Assume that $t > s$, $(t, s \in I^*)$. Then

$$E\{N_i(t)|\mathbf{B}_s(X)\} = M_i(s) - E\left\{E\left(M_i(t)|\bigvee_{j\le i-1}\mathbf{B}_t(M_j)\right)|\mathbf{B}_s(X)\right\} \tag{4.4}$$

holds. Since $\bigvee_{j\le i-1}\mathbf{B}_t(N_j)$ and $\bigvee_{j\ge i}\mathbf{B}_s(N_j)$ are proved to be independent, we have

$$\begin{aligned}\text{the right-hand side of (4.4)} &= M_i(s) - E\left\{M_i(t)|\bigvee_{j\le i-1}\mathbf{B}_s(M_j)\right\}\\ &= N_i(s),\end{aligned}$$

with probability 1.

(iii) Using the system of martingales N_i, $i = 1, 2, \ldots$, we set

$$A_i(t) = E(N_i(t)^2)/E(N_i(\tau_i)^2), \qquad t \in I^*.$$

Then $A_i(t)$ is an increasing function on I^* and the number of t's such that either $A_i(t-) \ne A_i(t)$ or $A_i(t) \ne A_i(t+)$ holds is countable. In addition, we have

$$M_i(t) = M_i(\tau_i), \qquad t \ge \tau_i,$$

which comes from the definition. Hence,

$$N_i(t) = N_i(\tau_i), \qquad t \ge \tau_i.$$

With this remark and with the inequality $A_i(t) \le 1$, we see that $A_i(t) = 1$ for $t \ge \tau_i$. By using this family $\{A_i;\ i = 1, 2, \ldots\}$ of increasing functions we can form a system $\{m_i^*(dt);\ i = 1, 2, \ldots\}$ of Stieltjes measures on I^* as is shown below. Eventually, it will lead to the system of spectral measures.

Now set $A(t) = \sum_{i=1}^{\infty} A_i(t)/2^i$. Then, $A(t)$ is an increasing function and satisfies the inequality $A(t) \le 1$. In addition, $dA_i(t)$ is absolutely continuous with respect to $dA(t)$ for every i; $dA_i(t) \ll dA(t)$. Hence, by the Radon-Nikodym Theorem, there exists a density function $\alpha_i(t)$ such that $dA_i(t) = \alpha_i(t)\, dA(t)$. Set

$$\Delta_j = \{t \in I^*;\ \alpha_j(t) \ne 0\}.$$

A family $\{\Delta_{ij};\ i \le j\}$ of double indexed Borel sets is defined in the following manner:

$$\Delta_{11} = \Delta_1, \qquad \Delta_{1j} = \Delta_j - \bigcup_{r=1}^{j-1}\Delta_r, \qquad j \ge 2,$$

$$\Delta_{jj} = \Delta_j - \bigcup_{\mu=1}^{j-1}\Delta_{\mu j},$$

$$\Delta_{ij} = \Delta_j - \bigcup_{\mu=1}^{i-1}\Delta_{\mu j} - \bigcup_{\gamma=i}^{j-1}\Delta_{i_\gamma}, \qquad i < j.$$

By definition, it is easy to see that for each fixed j, the sets Δ_{ij}, $i = 1, 2, \ldots, j$, are disjoint and their sum is Δ_j.

We now define the measures $m_i^*(dt)$:

$$m_i^*(dt) = \sum_{j=1}^{\infty} 2^{-j} \chi_{ij}(t)\, dA_j(t), \qquad \chi_{ij}(t) = \begin{cases} 1, & t \in \Delta_{ij}, \\ 0, & \text{otherwise}. \end{cases} \tag{4.5}$$

Obviously, $m_{i-1}^*(dt) \gg m_i^*(dt)$ and $dA(t) \gg m_i^*(dt)$, $i = 1, 2, \ldots$, hold. Let m_i^* be decomposed into the continuous part m_i and the discontinuous parts m_{ij}^-, m_{ij}^+:

$$m_i^*(dt) = m_i(dt) + \sum_j m_{ij}^- \delta_{t_j -}(dt) + \sum_j m_{ij}^+ \delta_{t_j +}(dt), \tag{4.6}$$

where $\delta_{t_j \pm}(dt) = 1$, if $t_j \pm \in dt$, and $= 0$ otherwise. We see that $m_{i-1}(dt) \gg m_i(dt)$ holds for every i and that $m_{i-1,j}^{\pm} = 0$ implies $m_{i,j}^{\pm} = 0$. Note that $m_j(dt)$ is a continuous measure on I^*; however, it may be viewed as a continuous measure on I.

(iv) We are now ready to construct an innovation process (which is a Gaussian martingale) $B = \{B(t);\ t \in I^*\}$ from the system $\{N_i;\ i = 1, 2, \ldots\}$ of martingales in order to form the canonical representation. Taking the family of functions χ_{ij} used in the formula (4.5) we set

$$B_i^*(t) = \sum_{j \geq i} \chi_{ij}(s) \{2^j E[N_j(\tau_j)^2]\}^{-1/2}\, dN_j(s).$$

Then we have

$$\begin{aligned} E[B_j^*(t)^2] &= \int_0^t \sum_{j \geq i} \chi_{ij}(s) \{2^j E[N_j(\tau_j)^2]^{-1} E(dN_j(s))^2 \\ &= \int_0^t \sum_{j \geq i} 2^{-j} \chi_{ij}(s)\, dA_j(s) = m_i^*[0, t], \qquad t \in I^*. \end{aligned}$$

Since the N_{ij}, $j \geq i$, $i = 1, 2, \ldots$, with $N_{ij}(t) = \int_0^t \chi_{ij}(s)\, dN_j(s)$ are independent, we see that if $i \neq j$, then

$$E[B_j^*(t) B_j^*(s)] = 0, \qquad \text{for any } t, s \in I^*;$$

that is B_i and B_j are independent. Also, we have, by definition, the following equality

$$\mathbf{B}_t(N_j) = \bigvee_{i \leq j} \mathbf{B}_t(N_{ij}) \quad \text{for any } t \in I^*,$$

where $\mathbf{B}_t(N_{ij}) = \sigma\{N_{ij}(s);\ s \leq t\}$. Together with the σ-field $\mathbf{B}_t(B_i^*)$, we can present the following relations:

$$\bigvee_j \mathbf{B}_t(N_j) = \bigvee_j \bigvee_{i \leq j} \mathbf{B}_t(N_{ij}) = \bigvee_i \bigvee_{j \geq i} \mathbf{B}_t(N_{ij}) = \bigvee_i \mathbf{B}_t(B_i^*).$$

Summing up, the equality

$$\bigvee_i \mathbf{B}_t(B_i^*) = \mathbf{B}_t(X) \quad \text{for } t \in I^*$$

has now been proved.

A decomposition of $B_i^*(t)$ into independent terms is obtained in accordance with the decomposition (4.6):

$$B_i^*(t) = B_i(t) + \sum_j (m_{ij}^-)^{1/2} B_{t_j-}^i + \sum_j (m_{ij}^+)^{1/2} B_{t_j+}^i, \qquad t \in I^*.$$

By the remark given at the end of (iii), $B_i = \{B_i(t);\ t \in I^*\}$ may be identified with $B_i = \{B_i(t);\ t \in I\}$, by taking I instead of I^*. It is easy to see that

$$\{B_i,\ i = 1, 2, \ldots;\ B_{t_j-}^i,\ i, j = 1, 2, \ldots;\ B_{t_j+}^i,\ i, j = 1, 2, \ldots\}$$

is a system of independent random variables, where all the $B_{t_j-}^i$ and the $B_{t_j+}^i$ are subject to the standard Gaussian distribution $N(0, 1)$. The decomposition (4.6) claims that the set t_j is countable. Take a t_j and rearrange $\{B_{t_j-}^i, B_{t_j+}^i;\ i = 1, 2, \ldots\}$ to denote $\{B_j^l;\ l = 1, 2, \ldots, L_j\}$. Set

$$B_j^l(t) = \begin{cases} 0 & \text{if } t < t_j, \\ B_j^l & \text{if } t \geq t_j, \end{cases} \qquad \text{in the case } B_j^l = B_{t_j-}^j,$$

$$B_j^l(t) = \begin{cases} 0 & \text{if } t \leq t_j, \\ B_j^l & \text{if } t > t_j, \end{cases} \qquad \text{in the case } B_j^l = B_{t_j+}^j.$$

By using the B_i and B_j^l we define $B = \{B(t);\ t \in I\}$ by (4.1).

(v) We will prove that B is an innovation process. It is almost obvious that B satisfies condition (I.1), (I.2) and (I.3). We now prove that (I.4) is satisfied. To this end a general remark is given. Let $f(\omega)$ be a $\mathbf{B}_s(X)$-measurable random variable and let $\{X(t);\ t \in T\} \cup \{f\}$ be a Gaussian system. Set

$$L(t) = E[f|\mathbf{B}_t(X)], \qquad t \in T^*,\ t \leq s,$$

Then $L(t)$ is a Gaussian martingale and the system $\{X(t);\ t \in T\} \bigcup \{L(t);\ t \in T^*\}$ is again Gaussian (see §2.2). Noting that $\bigvee_j \mathbf{B}_t(B_j^*) = \mathbf{B}_t(X)$, we have

$$f = \sum_{i=1}^N \int_0^s F_i(u)\, dB_i(u) + \sum_{\{t_j \leq s\}} \sum_{i=1}^{L_j} b_j^l B_j^l(t) \quad \text{(a.s. } P\text{)},$$

where the kernels $F_j(u)$ are measurable functions of u and then b_j^l are constants such that

$$\sum_{j=1}^N \int_0^s F_i(u)^2\, m_i(du) + \sum_{\{t_j \leq s\}} \sum_{i=1}^{L_j} (b_j^l)^2 < \infty.$$

If, in particular, $f = X(t)$, then $F_i(u) = F_i(t, u)$ and $b_j^l = b_j^l(t)$ hold for every t, and the representation (4.3) is obtained. The equality $\mathbf{B}_t(X) = \mathbf{B}_t(B)$ is now obvious.

(vi) Finally, we prove assertion (II) of the theorem. First, we note the relations $\mathbf{B}_t(B) = \mathbf{B}_t(X) = \mathbf{B}_t(\widetilde{B})$, where

$$\begin{aligned}\widetilde{B}(t) &= (\widetilde{B}_1(t), \dots, \widetilde{B}_n(t), \dots ; \widetilde{B}_1^1(t), \dots, \widetilde{B}_1^l(t), \dots ; \widetilde{B}_j^1(t), \dots, \widetilde{B}_j^l(t), \dots) \\ &= (\widetilde{B}_i(t),\ i = 1, \dots, \widetilde{N};\ \widetilde{B}_j^l(t),\ j = 1, 2, \dots,\ l = 1, \dots, \widetilde{L}_j)\end{aligned}$$

is the innovation process of X for the representation (4.3$'$). First we show that two measures m_i and $\tilde{m}_i$ ($\tilde{m}_i(du) = E(|d\widetilde{B}_i(u)|^2)$) are equivalent. Since $B_i = \{B_i(t);\ t \in I\}$ is a continuous martingale with respect to $\mathbf{B}_t(\widetilde{B})$ ($= \mathbf{B}_t(B)$), there is no discontinuous martingale $B_j^l = \{B_j^l(t);\ t \in I\}$ in the expression of $B_i(t)$. Thus, we are given

$$B_i(t) = \sum_{k=1}^{\widetilde{N}} \int_0^t \alpha_{ik}(u)\, d\widetilde{B}_k(u), \qquad \sum_{k=1}^{\widetilde{N}} \int_0^t \alpha_{ik}(u)^2\, \tilde{m}(du) < \infty, \qquad t \in I.$$

In this situation, the following equality

$$m_i(du) = \sum_{k=1}^{\widetilde{N}} \alpha_{ik}(u)^2\, \tilde{m}_k(du)$$

must hold. Hence, noting the relations $\tilde{m}_k(du) \ll \tilde{m}_1(du)$ for $k \geq 2$, we conclude $m_i(du) \ll m_1(du)$, in particular $m_1(du) \ll \tilde{m}_1(du)$. A similar argument guarantees the opposite relation $\tilde{m}_1(du) \ll m_1(du)$. Hence $\tilde{m}_1$ and m_1 are equivalent.

As for equivalence in the case of general n, we prove this by induction. Assume the relation $m_i(du) \sim \tilde{m}_i(du)$ holds for every $i \leq n$. Since $\tilde{m}_{n+1}(du) \ll \tilde{m}_n(du)$, there exists a density function $\beta(u)$ such that $\tilde{m}_{n+1}(du) = \beta(u)\tilde{m}_n(du)$ (The Radon-Nikodym Theorem). Set $A = \{u;\ \beta(u) = 0\}$. To prove $m_{n+1}(du) \ll \tilde{m}_{n+1}(du)$ it suffices for us to show that $m_{n+1}(A) = 0$. The condition $m_{n+1}(A) = 0$ is equivalent to the assertion that the Gaussian martingale $M_{n+1}(A, t)$ given by

$$M_{n+1}(A, t) = \int_0^t \chi_A(u)\, dB_{n+1}(u) = \sum_{k=1}^{\widetilde{N}} \int_0^t \chi_A(u) \alpha_{n+1,k}(u)\, d\widetilde{B}_k(u) \tag{4.7}$$

vanishes with probability one. We now proceed to prove the last property.

Since $\tilde{m}_i(A) = 0$ for $i \geq n+1$, the right-hand side of (4.7) is the sum of the terms up to the nth one

$$\sum_{k=1}^{n} \int_0^t \chi_A(u) \alpha_{n+1,k}(u)\, d\widetilde{B}_k(u).$$

While

$$M_i(A, t) = \int_0^t \chi_A(u)\, dB_i(u) = \sum_{k=1}^{\widetilde{N}} \int_0^t \chi_A(u)\alpha_{ik}(u)\, d\widetilde{B}_k(u)$$

$$= \sum_{k=1}^{n} \int_0^t \chi_A(u)\alpha_{ik}(u)\, d\widetilde{B}_k(u), \qquad i = 1, 2, \dots, n, \tag{4.8}$$

are mutually independent martingales, so that

$$E[M_i(t)M_j(t)] = \int_0^t \sum_{k=1}^{n} \alpha_{ik}(u)\alpha_{jk}(u)\chi_A(u)\tilde{m}_k(du) = 0, \qquad t \in I$$

holds for every pair (i, j) with $i \neq j$. (Namely, the vectors $\alpha_i(u) = (\alpha_{i1}(u), \dots, \alpha_{in}(u))$, $i = 1, \dots, n$, are orthogonal to the Hilbert space $L^2(\tilde{m}_1, \tilde{m}_2, \dots, \tilde{m}_n) = \{(f_1, f_2, \dots, f_n);\ f_i \in L^2(A, \tilde{m}_i)\}$ of vector-valued functions.) On the other hand, the vectors $\alpha_i(u)$, $i = 1, 2, \dots, n$, should be linearly independent on A for $\tilde{m}_n$-almost all u. In fact, assume that there were a subset D of A such that $m_n(D) > 0$ and that

$$\alpha_j(u) = \sum_{i \neq j} \gamma_i(u)\alpha_i(u) \quad \text{on } D,$$

holds for some j. Then the martingale

$$M_j(D, t) = \sum_{k=1}^{\widetilde{N}} \int_0^t \chi_D(u)\alpha_{jk}(u)\, d\widetilde{B}_k(u) = \sum_{k=1}^{n} \int_0^t \chi_D(u)\alpha_{jk}(u)\, d\widetilde{B}_k(u)$$

should be equal to a martingale of the form

$$\sum_{i \neq j} \int_0^t \gamma_i(u)\, dM_i(D, u)$$

with probability one for any t. This fact contradicts the fact that the martingales $M_1(D, 1), \dots, M_n(D, t)$ are independent.

Since martingales M_{n+1} and M_i $(i = 1, \dots, n)$ are mutually independent, we must have

$$E[M_{n+1}(A, t)M_i(A, t)] = \sum_{i=1}^{n} \int_0^t \chi_A(u)\alpha_{n+1,k}(u)\alpha_{i,k}(u)\tilde{m}_k(du) = 0.$$

Noting that $\alpha_i(u)$, $i = 1, \dots, n$, are linearly independent, we see that

$$\alpha_{n+1}(u) = (\alpha_{n+1,1}(u), \dots, \alpha_{n+1,n}(u))$$

is zero for $\tilde{m}_n$-almost everywhere on A. Hence, $m_{n+1}(A) = 0$.

In the same manner it can be proved that $\tilde{m}_{n+1}$ is absolutely continuous with respect to m_{n+1}. Hence, the equivalence $m_i \sim \tilde{m}_i$ for every i $(i \leq N)$ has been proved. At the same time the equality $N = \widetilde{N}$ is proved. (Thus, continuous multiplicity is well defined.)

As for the discrete spectrum, we can first prove that the two sets $\{t_1, t_2, \dots\}$ and $\{\tilde{t}_1, \tilde{t}_2, \dots\}$ are the same, and then we can easily show that $L_j = \widetilde{L}_k$ if $t_j = \tilde{t}_k$, so the details are omitted. □

Here is a short remark. In Theorem 4.1, the parameter set is taken to be $[0, \infty)$; however, the set can be restricted to be a finite interval, say $[0, 1]$. In the case where $I = (-\infty, \infty)$, we need a slight modification of the statement; the innovation process is $B = \{B(t);\ t \in (-\infty, \infty)\}$, the integral sign $\int_0^t$ in (4.3) has to be replaced with $\int_{-\infty}^t$ noting square itegrability of the kernels over $(-\infty, \infty)$, and so forth. With this remark, we now state

THEOREM 4.1′. (II) *Let $X = \{X(t);\ t \in (-\infty, \infty)\}$ be a separable Gaussian process. If X satisfies the purely nondeterministic property*

$$\bigcap_{t \subset (-\infty, \infty)} \mathbf{B}_t(X) = \{\varnothing, \Omega\}, \tag{4.9}$$

then X has the canonical representation of the form

$$X(t) = \sum_{i=1}^{N} \int_{-\infty}^{t} F_i(t, u)\, dB_i(u) + \sum_{\{t_j \le t\}} \sum_{l=1}^{L_j} b_j^l(t) B_j^l(t), \tag{4.10}$$

where

$$\sum_{i=1}^{N} \int_{-\infty}^{t} F_i(t, u)^2 m_i(du) < \infty \quad \text{and} \quad \sum_{\{t_j \le t\}} \sum_{l=1}^{L_j} b_j^l(t)^2 < \infty, \qquad t \in (-\infty, \infty).$$

(II) *If X has another canonical representation*

$$X(t) = \sum_{i=1}^{\widetilde{N}} \int_{-\infty}^{t} \widetilde{F}_i(t, u)\, d\widetilde{B}_i(u) + \sum_{\{t_j' \le t\}} \sum_{l=1}^{\widetilde{L}_j} b_j^l(t) \widetilde{B}_j^l, \tag{4.10′}$$

then $N = \widetilde{N}$ and $m_i \sim \tilde{m}_i$ with $\tilde{m}_i(du) = E[|d\widetilde{B}_i(u)|^2]$, $i = 1, 2, \dots$, and $\{t_1, t_2, \dots\} = \{t_1', t_2', \dots\}$ hold. It also holds that $L_i = \widetilde{L}_k$ if $t_i = t_k$.

PROOF. Change the parameter, by setting $\tau = e^t$, to define $Y(\tau) = X(t)$. Tacitly, assuming that $Y(0) = 0$, we are given a Gaussian process $Y = \{Y(\tau);\ \tau \in [0, \infty)\}$. The assumption (4.9) guarantees the condition $\bigcap_{\tau > 0} \mathbf{B}_\tau = \{\phi, \Omega\}$. Therefore, we can appeal to Theorem 4.1 to obtain the canonical representation of Y:

$$Y(\tau) = \sum_{i=1}^{N} \int_{0}^{t} G(\tau, v)\, d\overline{B}_i(v) + \sum_{\{\tau_i \le \tau\}} \sum_{l=1}^{L_i} a_i^l(\tau) \overline{B}_i^l. \tag{4.11}$$

The right-hand side should be expressed so as to include only nonzero terms. Returning to the original parameter $t : t = \log \tau$, $\tau > 0$, and forming $B_i(t)$ and $B_i^l(t)$ from $\overline{B}_i(\tau)$ and $\overline{B}_i^l(\tau)$, respectively, we obtain the representation (4.10) from (4.11). Assertion (II) can be proved as in Theorem 4.1. □

REMARK. The Karhunen representation, discussed in §3.4, of a purely non-deterministic mean continuous stationary process is a particular, and in fact typical, case where $N = 1$ and no discrete spectrum arises.

§3. Gaussian processes and reproducing kernel Hilbert spaces

Let $X = \{X(t);\ t \in I\}$ be a Gaussian process with $E(X(t)) \equiv 0$, and take the covariance function $\Gamma(t, s) = E(X(t)X(s))$. Since $\Gamma(t, s)$ is non-negative definite, there exists a reproducing kernel Hilbert space $\mathbf{H}$ with (reproducing) kernel $\Gamma(t, s)$. That is, $\mathbf{H}$ is a vector space involving functions $f = (f(t);\ t \in I)$ satisfying the conditions (R.1) and (R.2):

(R.1) $\Gamma(\cdot, s) \in \mathbf{H}$ for every $s \in I$,

(R.2) (Reproducing property) For any $f \in \mathbf{H}$ and for any $s \in I$, $\langle f(\cdot), \Gamma(\cdot, s)\rangle = f(s)$, where $\langle\ ,\ \rangle$ denotes the inner product in $\mathbf{H}$.

(R.3) $\mathbf{H}$ is generated by the family $\{\Gamma(\cdot, s);\ s \in I\}$

REMARK. If we take $f(\cdot) = \Gamma(\cdot, t)$, then the equation (R.2) implies the interesting relation

$$\langle \Gamma(\cdot, s), \Gamma(\cdot, t)\rangle = \Gamma(t, s),$$

where Γ looks like a factor of itself.

In what follows in this section, we shall discuss several basic relations between the reproducing kernel Hilbert space $\mathbf{H}$ with kernel $\Gamma(t, s)$, which is the covariance function of a Gaussian process X, and its canonical representation. For this purpose we need to introduce some notation. Let $\mathbf{H}_t$ be the subspace of $\mathbf{H}$ spanned by the family $\{\Gamma(\cdot, \tau);\ \tau < t\}$. Similarly $\mathbf{H}_{t-}$ is defined by the family $\{\Gamma(\cdot, \tau);\ \tau < t\}$. Set $\mathbf{H}_{t+} = \bigcap_{\tau > t} \mathbf{H}_\tau$.

Another space to be introduced is a Hilbert space $H(X)$ spanned, in the space $L^2(\Omega, P)$, by the Gaussian process $X = \{X(t);\ t \in I\}$ defined on $(\Omega, \mathbf{B}, P)$. Take a vector space

$$\overset{\circ}{H}(X) = \left\{\sum a_i X(t_i) \text{ finite sum; } a_i \text{ real, } t_i \in I\right\} \tag{4.12}$$

and have it completed with respect to the $L^2(\Omega, P)$-norm to get the required space $H(X)$. In other words, $H(X)$ is the closed linear manifold spanned by the $X(t)$, $t \in I$. The inner product $(\ ,\ )$ in $H(X)$ is given by the covariance: $(X_1, X_2) = E(X_1 X_2)$ because of the assumption (A.1) $E(X(t)) = 0$. Restrict the values of the t_j in the $\{\ \}$ of (4.12) to the region $[0, t]$ $(\subset I)$ to define a vector space

$$\overset{\circ}{H}_t(X) = \left\{\sum a_i X(t_i) \text{ finite sum; } a_i \text{ real}, t_i \in [0, t]\right\}.$$

Then a subspace $H_t(X)$ of $H(X)$ is defined by taking a closure. The subspace $H_t(X)$ is often called the *linear manifold spanned by the values of* X *before the instant* t. The subspace $H_{t-}(X)$ is the completion of $\bigcup_{s<t} H_s(X)$

and $H_{t+}(X)$ is $\bigcap_{s>t} H_s(X)$. These subspaces $H_{t-}(X)$ and $H_{t+}(X)$ are increasing in $t \in I^*$ (see §2.); more precisely, for $t > s$, we have

$$H_s(X) \subset H_{s+}(X) \subset H_{t-}(X) \subset H_t(X).$$

We shall establish a remarkable isomorphism between the reproducing kernel Hilbert spaces and linear manifolds spanned by the X as in the next theorem.

THEOREM 4.2. *Let $\Gamma(t, s)$ be the covariance function of $X = \{X(t);\ t \in I\}$, and let S be the mapping that maps $\Gamma(\cdot, t) \in \mathbf{H}$ to $X(t) \in H(X)$. Then S extends to a linear isomorphism between $\mathbf{H}$ and $H(X)$.*

PROOF. Let S extend to a mapping from $f(s) = \sum_{i=1}^n a_i \Gamma(s, t_i)$ in $\mathbf{H}$ to $X_f = \sum_{i=1}^n a_i X(t_i) \in H(X)$: $Sf = X_f$. By using the relation

$$\langle \Gamma(\cdot, t), \Gamma(\cdot, s) \rangle = \Gamma(t, s) = E(X(t)X(s)),$$

we see easily that $\langle f, g \rangle = E(X_f X_g)$, where $g(s) = \sum_{j=1}^m b_j \Gamma(s, t_j)$ and $X_g = Sg = \sum_{j=1}^m b_j X(t_j)$. By isometry, S can be extended to a unitary mapping with domain $\mathbf{H}$ and range $H(X)$. □

By the reproducing property of f in $\mathbf{H}$ and by the isometry S, we have

COROLLARY 4.1. *Every element f in $\mathbf{H}$ is expressed in the form*

$$f(t) = E(X(t)X_f) \quad \text{with } X_f = Sf \in H(X).$$

PROOF. Use the reproducing property $f(t) = \langle \Gamma(\cdot, t), f(\cdot) \rangle$ to prove the assertion. □

COROLLARY 4.2. *The mapping S carries $\mathbf{H}_t$, $\mathbf{H}_{t+}$, and $\mathbf{H}_{t-}$ to $H_t(X)$, $H_{t+}(X)$, and $H_{t-}(X)$, respectively.*

The proof is obvious and is thus omitted.

We now return to the canonical representation (4.3) of the Gaussian process $X = \{X(t);\ t \in I\}$. For the innovation process $B = \{B(t);\ t \in I\}$ we define the Hilbert space $H(B)$ and its subspaces $H_t(B)$, $H_{t+}(B)$, and $H_{t-}(B)$ as follows:

$$H(B) = \left\{ x\ ;\ x = \sum_{i=1}^N \int_0^\infty \alpha_i(u)\, dB_i(u) + \sum_j \sum_{l=1}^{L_j} \beta_j^l B_j^l, \right. \tag{4.13}$$

$$\left. \sum_{i=1}^N \int_0^\infty \alpha_i(u)^2\, m_i(du) + \sum_j \sum_{l=1}^{L_j} (\beta_j^l)^2 < \infty \right\},$$

$$H_t(B) = \left\{ x \in H(B);\ x = \sum_{i=1}^N \int_0^t \alpha_i(u)\, dB_i(u) + \sum_{\{t_j \le t\}} \sum_{l=1}^{L_j} \beta_j^l B_j^l(t) \right\},$$

and $H_{t+}(B) = \bigcap_{s>t} H_s(B)$, $H_{t-}(B) =$ closure of $\bigcup_{s<t} H_s(B)$. Since the canonical representation (4.3) can be formed only by linear operations, we

can see the following equalities:

$$\begin{aligned} H(B) = H(X), \qquad H_t(B) = H_t(X), \\ H_{t+}(B) = H_{t+}(X), \quad \text{and} \quad H_{t-}(B) = H_{t-}(X). \end{aligned} \tag{4.14}$$

REMARK. Take an element x in $H(X) = H(B)$. If x is written in the form $x = \sum_{i=1}^N \int_0^\infty \alpha_i(u)\, dB_i(u) + \sum_j \sum_{l=1}^{L_j} \beta_j^l B_j^l(t)$, then its conditional expectation $E[x|\mathbf{B}_t(X)]$ is

$$\sum_{i=1}^N \int_0^t \alpha_i(u)\, dB_i(u) + \sum_{\{t_j \le t\}} \sum_{i=1}^{L_j} \beta_j^l B_j^l(t),$$

which is the projection of x down to the subspace $H_t(X) = H_t(B)$. Similarly, the conditional expectation relative to the σ-field $\mathbf{B}_{t+}(X)$ and that to $\mathbf{B}_{t-}(X)$ are projections to the subspaces $H_{t+}(X) = H_{t+}(B)$ and $H_{t-}(X) = H_{t-}(B)$, respectively.

From what we have discussed we obtain a representation of the elements in a reproducing kernel Hilbert space.

THEOREM 4.3. *Let $\Gamma(t, s)$ be the covariance function of a Gaussian process having canonical representation* (4.3), *and let $\mathbf{H}$ be the reproducing kernel Hilbert space with kernel $\Gamma(t, s)$. Then, an element f of $\mathbf{H}$ is expressed in the form*

$$f(t) = \sum_{j=1}^N \int_0^t F_j(t, u)\alpha_j(u)\, m_j(du) + \sum_{\{t_j \le t\}} \sum_{l=1}^{L_j} b_j^l(t)\beta_j^l, \tag{4.15}$$

where

$$\sum_{j=1}^N \int_0^\infty \alpha_j(u)^2\, m_i(du) + \sum_j \sum_{i=1}^{L_j} (\beta_j^l)^2 < \infty.$$

The α_i, $1 \le i \le N$, and β_j^l, $j \ge 1$, $1 \le l \le L_j$, are uniquely determined by f.

PROOF. By Corollary 4.1 we have the expression

$$f(t) = E[X(t)x], \qquad x \in H(X).$$

Assume that x can be written in the form

$$x = \sum_{i=1}^N \int_0^\infty \alpha_i(u)\, dB_i(u) + \sum_j \sum_{i=1}^{L_j} \beta_j^l B_j^l, \tag{4.16}$$

and obtain the covariance of x and $X(t)$ to prove formula (4.15). For uniqueness it suffices for us to show that $f = 0$ implies the equalities $\alpha_i(t) \equiv 0$, $1 \le i \le N$, and $\beta_j^l = 0$, $j \ge 1$, and $1 \le l \le L_j$. Since $x = Sf = 0$, each member of the right-hand side of (4.16) has to be 0, i.e.,

$$\int_0^\infty \alpha_i(u)\, dB_i(u) = 0, \qquad 1 \le i \le N \quad \text{and} \quad \beta_j^l B_j^l = 0.$$

Hence, follows our assertion. □

A partial converse of Theorem 4.3 holds, and in fact, it gives us a criterion for the canonical property of a representation.

THEOREM 4.4. *Assume that a Gaussian process* $X = \{X(t);\ t \in I\}$ *is represented in the form* (4.3) *by an independent increment process* $B = \{B(t);\ t \in I\}$ *of the form* (4.1). *A necessary and sufficient condition for the representation to be canonical* (*in other words,* B *is the innovation process of* X) *is that if, for any* $T \in I$, *a function* $f(t)$ *given by* (4.15) *is identically* 0 *on* $[0, T]$, *then* $\alpha_j(t) = 0$ *for* $t \leq T$ *and* $\beta_j^l = 0$ (*for such* l, j *that* $t_j < T$ *or* $t_j = T$ *with* $b_j^l(T) \neq 0$) *hold.*

PROOF. Necessity is obvious from the second half of the proof of Theorem 4.3. Sufficiency is proved as follows. Suppose that there exist an α_i $(1 \leq i \leq N)$ and a β_j^l $(j \geq 1,\ 1 \leq l \leq L_j)$ that are not 0 and satisfy the equation

$$f(t) = \sum_{i=1}^{N} \int_0^t F_i(t, u)\alpha_i(u)\, m_i(du) + \sum_{\{t_j \leq t\}} \sum_{l=1}^{L_j} \beta_j^l b_j^l(t) = 0, \qquad t \leq T. \tag{4.17}$$

Then any y in $H_T(B)$ being expressed in the form

$$y = \sum_{i=1}^{N} \int_0^T \alpha_i(u)\, dB_i(u) + \sum_j \sum_{l=1}^{L_j} \beta_j^l B_j^l(T),$$

is orthogonal to $X(t)$ for every $t \in [0, T]$. This means that y is not in $H_t(X)$, $t \in [0, T]$, if $y \neq 0$. Hence, $H_t(X)$ is a proper subspace of $H_T(B)$. This contradicts to the assumption that (4.3) is canonical. □

Theorem 4.3 and Theorem 4.4 imply that the reproducing kernel Hilbert space $\mathbf{H}$ admits a direct sum decomposition of the form

$$\mathbf{H} = \sum_{i=1}^{N} \bigoplus \mathbf{H}(i) \oplus \sum_j \sum_{l=1}^{L_j} \bigoplus \mathbf{H}(j, l), \tag{4.18}$$

where

$$\mathbf{H}(i) = \left\{ f;\ f(t) = \int_0^t F_i(t, u)\alpha(u)\, m_i(du),\ \alpha \in L^2(m_i) \right\},$$

and

$$\mathbf{H}(j, l) = \{f;\ f(t) = b_j^l(t)\beta,\ \beta \in R\}.$$

Now, some significant properties of this decomposition are shown.

COROLLARY 4.3. *Let* P_t (*or* P_{t+} *or* P_{t-}), $P(i)$, *and* $P(j, l)$ *be the projection operators from* $\mathbf{H}$ *down to* $\mathbf{H}_t$ (*or* $\mathbf{H}_{t+}$ *or* $\mathbf{H}_{t-}$), $\mathbf{H}(i)$, *and* $\mathbf{H}(j, l)$, *respectively. Then the following commutativity relations hold*:

$$P(i)P_t = P_tP(i), \quad P(j,l)P_t = P_tP(j,l), \quad P(i)P_{t+} = P_{t+}P(i),$$
$$P(i)P_{t-} = P_{t-}P(i), \quad P(j,l)P_{t+} = P_{t+}P(j,l) \quad \text{and} \quad P(j,l)P_{t-} = P_{t-}P(j,l).$$

PROOF. The components of the innovation process **B** are mutually independent and they have independent increments. These properties can be rephrased in terms of reproducing kernel Hilbert spaces and the assertion is proved. □

COROLLARY 4.4. *The reproducing kernel Hilbert spaces* $\mathbf{H}(i)$ *and* $\mathbf{H}(j,l)$, *which are subspaces of* $\mathbf{H}$, *have reproducing kernels*

$$\Gamma_{(i)}(s,t) = \int_0^{t\wedge s} F_i(t,u)F_i(s,u)\, m_i(du) \quad \text{and} \quad \Gamma_{(j,l)}(s,t) = b_j^l(t)b_j^l(s),$$

respectively.

§4. Examples of a canonical representation and a noncanonical one

A. Brownian motion (Wiener process). Let $B = \{B(t);\ t \in [0,\infty)\}$ be a one-dimensional Brownian motion (Gaussian, $E(B(t)) \equiv 0$, $E(B(t)B(s)) = t \wedge s$), which is often called a standard Brownian motion. As was mentioned in the example in §2, B can be represented canonically by itself. It has unit multiplicity, and the spectral measure m_1 is of Lebesgue type $E\{|dB(u)|^2\} = du$. There is no discrete spectrum. The associated reproducing kernel Hilbert space is

$$\mathbf{H} = \left\{ f;\ f(t) = \int_0^t \alpha(u)\,du,\ \int_0^\infty \alpha(u)^2\,du \equiv \|f\|^2 < \infty \right\}.$$

B. It is of interest to give examples of noncanonical representations.

(i) Let $B_1 = \{B_1(t);\ t \in [0,\infty)\}$ and $B_2 = \{B_2(t);\ t \in [0,\infty)\}$ be mutually independent standard Brownian motions, and set

$$\overline{B}(t) = \frac{1}{\sqrt{2}}\{B_1(t) + B_2(t)\}, \qquad t \in [0,\infty). \tag{4.19}$$

Then $\overline{B}$ is again a standard Brownian motion and (4.19) is regarded as a representation of $\overline{B}$ by B_1 and B_2. As is easily seen from Theorem 4.1 or from Theorem 4.4, the representation (4.19) is not canonical: $H_t(\overline{B}) \neq H_t(B_1) \oplus H_t(B_2)$.

(ii) The next example is, in a sense, rather technical. Take a Brownian motion B and define $B_0(t)$ by

$$B_0(t) = \int_0^t \left\{ 3 - 12\frac{u}{t} + 10\left(\frac{u}{t}\right)^2 \right\} dB(u)$$

(P. Lévy 1957). Then the covariance functions of $B_0 = \{B_0(t);\ t \in [0,\infty)\}$ is $t \wedge s$, and hence we see that B_0 is a Brownian motion. On the other hand, there is a random variable $Y_1 = \int_0^T u\,dB(u)$ which is independent of $B_0(t)$ for all $t \le T$. To prove this fact, check that the covariance of Y_1 and $B_0(t)$ vanishes for every $t \le T$. Hence, we have shown that $H_t(B_0) \ne H_t(B)$, which means that the above expression is a *noncanonical* representation of B_0.

Incidentally, $Y_2 = \int_0^T u^2\,dB(u)$ is also independent of $B_0(t)$, for all $t \le T$.

More generally, we can find, for any n, a polynomial P_n such that

$$X(t) = \int_0^t P_n\left(\frac{u}{t}\right) dB(u)$$

is a Brownian motion. For $n \ge 1$, the above representation is surely non-canonical. In fact, for any fixed T

$$Y_k = \int_0^T u^k\,dB(u), \qquad k = 1, 2, \ldots, n,$$

are independent of $X(t)$, $t \le T$.

C. Next we obtain a Gaussian process through a slight modification of a Brownian motion. In Chapter VI, it will be characterized as a process equivalent to a Brownian motion. The process $X = \{X(t);\ t \in [0, \infty)\}$ in question is defined by

$$X(t) = B(t) - \int_0^t \int_0^s k(s, u)\,dB(u)\,ds, \tag{4.20}$$

where B is, as usual, a Brownian motion, and where $k(s, u)$ is an L^2-kernel. The formula (4.20) may be rephrased as

$$X(t) = \int_0^t \left\{1 - \int_u^t k(s, u)\,ds\right\} dB(u) \equiv \int_0^t F(t, u)\,dB(u).$$

To prove that (4.20) is the canonical representation, it suffices to show that $\int_0^t F(t, u)\alpha(u)\,du = 0$ for all $t \le T$ implies $\alpha(t) = 0$ for almost all $t \in [0, T]$ (by Theorem 4.4). Take T arbitrarily and fix it. Assume that for any $t \le T$

$$\int_0^t F(t, u)\alpha(u)\,du = 0, \qquad t \le T.$$

Since the left-hand side is equal to

$$\int_0^t \left(1 - \int_u^t k(s, u)\,ds\right)\alpha(u)\,du - \int_0^t \left\{\alpha(s) - \int_0^s k(s, u)\alpha(u)\,du\right\} ds,$$

this implies

$$\alpha(t) = \int_0^t k(t, u)\alpha(u)\,du. \tag{4.21}$$

The uniqueness of the solution of an integral equation of Volterra type implies that $\alpha(t) = 0$ almost everywhere, and hence $\alpha(t) = 0$ in the L^2-space. Thus, we have proved that (4.20) is canonical. □

D. Gaussian processes with higher multiplicity. Let $B = \{B(t) = (B_1(t), B_2(t), \dots, B_N(t));\ t \in [0, \infty)\}$ be an N-dimensional Brownian motion and let $F(t)$ be a function satisfying

(i) $F(t) \neq 0$ on $[0, \infty)$.

(ii) $F(t)$ is absolutely continuous in t,

and

(iii) the Radon-Nikodym derivative $F'(t)$ belongs to $L^1([a, \infty)) = \{f;\ \int_a^\infty |f(t)|\, dt < \infty\}$, but $F(t)$ itself is not an element of the space $L^2([a, b]) = \{f;\ \int_a^b |f(t)|^2\, dt < \infty\}$ for any finite interval $[a, b] \subset [0, \infty)$.

Note that there certainly exists a function $F(t)$ satisfying the above conditions (i), (ii), and (iii). Using such a function F we can form a Gaussian process $X = \{X(t);\ t \in [0, \infty)\}$ by

$$X(t) = B_1(t) + F(t)B_2(t) + \cdots + F(t)^{N-1}B_N(t). \tag{4.22}$$

For simplicity we assume that N is finite. We now show that the expression (4.22) defines a canonical representation and the continuous multiplicity is N and that each special measure is of Lebesgue type. To this end, we first prove that if the equality

$$a_1(s) + F(s)a_2(s) + \cdots + F(s)a_N(s) = 0, \qquad s \leq T, \tag{4.23}$$

hold, then we must have $a_1(t) = 0$ for almost all $t \leq T$ (cf. Theorem 4.4). What we are concerned with now are functions $a_i(s)$ of the form $a_i(s) = \int_0^s \alpha_i(u)\, du$ with $\alpha_i \in L^2([0, \infty))$, so that each function $F(s)^{i-1}a_i(s)$ is absolutely continuous. Taking the Radon-Nikodym derivative, we have

$$\alpha_1(s) + F(s)\alpha_2(s) + \cdots + F(s)^{N-1}\alpha_N(s) + F'(s)\{a_2(s) + \cdots + (N-1)F(s)^{N-1}a_N(s)\} = 0, \quad \text{a.e.} \tag{4.24}$$

By assumption, $F'(s)$ is not square integrable on any finite interval, and hence

$$a_2(s) + \cdots + (N-1)F(s)^{N-1}a_N(s) = 0, \qquad \text{a.e., } s \leq T,$$

must be valid for (4.24) to hold. In fact, since the first term of the left-hand side of (4.24) is square integrable, so is the second term $F'(s)\{\,\cdot\,\}$. This situation can occur only when $\{\,\cdot\,\} = 0$. Repeating this argument, we can prove successively that each $a_i(s)$ has to be 0.

In the case $N = \infty$, we can still form a similar example. Again, by using the function $F(s)$ satisfying (i), (ii), and (iii), we can define $X = \{X(t);\ t \in [0, \infty)\}$ by

$$X(t) = \sum_{i=1}^{\infty} F(t)^{i-1}B_i(t), \tag{4.25}$$

where the B_i, $i = 1, 2, \dots$, are independent Brownian motions and where F is formed so as to satisfy $|F(s)| < 1$, for $s \in [0, \infty)$.

There is a remark concerning the above examples with higher multiplicity. There is a Gaussian process with higher multiplicity regardless how smooth it is. The sample function of X given by the formula (4.22) is continuous, as is seen from the continuity of Brownian motions B_i, $i = 1, 2, \ldots, N$. Define $Y = \{Y(t);\ t \in [0, \infty)\}$ by the formula $Y(t) = \int_0^t X(s)\,ds$. This process Y has, like X, a canonical representation using the same Brownian motion $B = \{B(t) = (B_1(t), \ldots, B_n(t));\ t \in [0, \infty)\}$. Repeating the integration as many times as we wish, we can form a Gaussian process that is smooth enough and is of multiplicity N. We therefore conclude that there exists a Gaussian process with arbitrarily high multiplicity even if the covariance is sufficiently smooth.

Incidentally, a Gaussian process with higher discrete multiplicity can also be formed without difficulty, and in fact it is easy to to construct a process with assigned discrete multiplicity.

E. Gaussian processes derived from the Lévy Brownian motion (Si Si's example). First we give the definition of the Lévy Brownian motion.

DEFINITION 4.4. A Gaussian system $X = \{X(A) = X(A, \omega);\ A \in R^N\}$ is called the *Lévy Brownian motion with* R^N-parameter if the following conditions (i), (ii), and (iii) are satisfied:

(i) $E(X(A)) = 0$, $A \in R^N$,

(ii) $X(O) = 0$, O being the origin of R^N,

(iii) $E\{X(A) - X(B)\}^2 = r(A, B)$, $r(A, B)$ being the distance between A and B in R^N.

The covariance function $\Gamma(A, B) = E(X(A)X(B))$ is easily computed and is given by

$$\Gamma(A, B) = \frac{1}{2}\{r(O, A) + r(O, B) - r(A, B)\}.$$

The existence of a Gaussian system with this covariance function is guaranteed by the theorem of I. J. Schönberg and L. Schwartz (see P. Lévy 1948, Chapter 8.)

The Si Si's example is obtained by restricting the parameter A to the unit circle passing through the origin. Then we have a one-dimensional parameter Gaussian process indexed by arc length θ. Keeping the symbol X, we denote the Gaussian process obtained in this manner by $X = \{X(\theta);\ \theta \in [0, 2\pi)\}$. The covariance function $\Gamma(\theta, \theta')$ is given by

$$\Gamma(\theta, \theta') = \sin(\theta/2)\{1 - \cos(\theta'/2)\} + \{1 + \cos(\theta/2)\}\sin(\theta'/2).$$

The canonical representation of $\{X(\theta)\}$ has been obtained (Si Si, 1987) as follows:

$$X(\theta) = \int_0^\theta \left\{ \sin(\theta/2)\left[\operatorname{cosec}(\theta'/2) - \frac{\cot(\theta'/4)}{2} h(\theta')\right] + \cos^2(\theta/4) h(\theta') \right\} dB(\theta'),$$

where $h(\theta) = \{1 + (\theta/4)\tan(\theta/4)\}^{-1}$.

REMARK. From the above expression of the canonical representation we see that the process $\{X(\theta)\}$ is a good example of a double Markov Gaussian process, the definition of which will be given in the next chapter.

If the given curve, to which the parameter is restricted, is not necessarily a circle but a $\mathbf{C}^3$-class curve, then it can be proved that a Gaussian process obtained from the Lévy Brownian motion has a canonical representation. The proof uses the fact that the curve C can be locally approximated by an arc of the osculating circle and the result mentioned above in the case of a circle. (For details see Si Si 1987.

§5. Application to prediction theory

As a good application of canonical representation theory, we shall briefly discuss the prediction theory for Gaussian processes.

First, an intuitive interpretation is given. Let a stochastic process $X = \{X(t)\}$ (not necessarily a Gaussian process) be given, and suppose that its values $X(u)$, $u \le s$, up to the instant s are observed, Then, prediction theory discusses the problem of how to estimate or forecast the future (unknown) value $X(t)$ $(t > s)$ as accurately as possible. To make matters concrete, it is necessary to introduce a measurement of the possible errors in order to evaluate the estimate. The mean square is the most popular and convenient measurement of the errors for our purpose. We therefore have to assume that the $X(t)$ has finite variance. The estimator itself is naturally $\mathbf{B}_s(X)$-measurable.

Incidentally, the expectation $E(X(t))$ can be assumed to be zero, since we are not interested in any nonrandom part of the process.

With the above discussion in mind, we give a setting for a mathematical theory of prediction. Given a stochastic process $X = \{X(t);\ t \in I\}$, I being an interval, satisfying the following conditions:

$$E(X(t)) = 0, \qquad E(X(t)^2) < \infty, \qquad t \in I. \tag{4.26}$$

For $X(t)$, t being given, we are in search of a $\mathbf{B}_s(X)$-measurable (with $s < t$) random variable Y with finite variance that minimizes the quantity

$$\sigma(t, s)^2 = E\{|X(t) - Y|^2\}. \tag{4.27}$$

The aim of *prediction theory* for the process X is to obtain such a random variable Y, which is called the *predictor* for $X(t)$ based on $\{X(u);\ u \le s\}$. The $\sigma(t, s)^2$ (≥ 0) given by (4.27) is called the *prediction error*. The predictor that minimizes $\sigma(t, s)^2$ is called the *best predictor* (relative to the mean square error).

The existence of the best predictor is guaranteed by

PROPOSITION 4.1. *Assume that X satisfies the assumption* (4.26). *For any t, s with $t > s$, the best predictor exists uniquely and it is equal to the conditional expectation $E(X(t)|\mathbf{B}_s(X))$.*

PROOF. The prediction problem in our setting can be rephrased in terms of the Hilbert space $L^2 = L^2(\Omega, \mathbf{B}, P)$, where $\mathbf{B}$ is taken to be $\sigma\{X(t); t \in I\}$. Set $L_s^2 = L^2(\Omega, \mathbf{B}_s(X), P)$. Then the $X(t)$ is a member of L_t^2, while Y is chosen from L_s^2. The prediction error is the distance in L^2 from $X(t)$ to the subspace L_s^2, so that the Y is uniquely determined as the orthogonal projection of $X(t)$ down to L_s^2. We now show that this projection is nothing but the conditional expectation $E(X(t)|\mathbf{B}_s(X))$. The inner product of $X(t) - E(X(t)|\mathbf{B}_s)$ and Y in L_s^2 is

$$\begin{aligned} E[\{(X(t) - E(X(t)|\mathbf{B}_s)\} \cdot Y] &= E(X(t)Y) - E\{E(X(t)|\mathbf{B}_s) \cdot Y\} \\ &= E(X(t)Y) - E\{E(X(t) \cdot Y|\mathbf{B}_s)\} \\ &= E(X(t)Y) - E\{X(t)Y\} \\ &= 0 \qquad \text{(since } Y \text{ is } \mathbf{B}_s\text{-measurable)} \end{aligned}$$

which was to be proved. □

We now come to our favorite case where the processes in question are Gaussian. By the definition of the canonical representation, we can easily prove the following assertion.

PROPOSITION 4.2. *If a Gaussian process $X = \{X(t);\ t \in I\}$ has a canonical representation in terms of $F(t, u)$ and $dB(u)$, then the best predictor $\widehat{X}(t, s)$ for $X(t)$ based on $\{X(u);\ u \le s\}$ with $s < t$ is given by*

$$\widehat{X}(t, s) = \int^s F(t, u)\, dB(u) \tag{4.28}$$

and the prediction error is

$$\sigma(t, s)^2 = \int_s^t F(t, u)^2\, m(du), \tag{4.29}$$

where $m(du) = E(|dB(u)|^2)$.

Thus, we have seen that the best predictor for a Gaussian process can immediately be obtained from its canonical representation. There is, however, an important remark; namely, the $\widehat{x}(t, s)$ in (4.28) is an element of the space $H_s(X)$ (in the notation established in §3). In other words, the best predictor is obtained as a *linear* functional of the $X(u)$, $u \le s$. This is a particularly important property of a Gaussian process which is worth mentioning.

Returning to the general case, the best predictor $\widehat{X}(t, s)$ is in general a nonlinear functional of the $X(u)$; $u \le s$, unlike the Gaussian case. It is therefore not easy to give an exact expression for the best predictor or to show how to compute it, and only a few special cases have so far been dealt with. Presently, our interest lies in linear prediction theory, where the best predictor is chosen from the linear manifold spanned by the $X(u)$, $u \le s$, that is from $H_s(X)$, which is a proper subspace of L_s^2. We therefore assume

(4.26) as before, and we wish to find Y in $H_s(X)$ that minimizes

$$E\{|X(t) - Y|^2\}. \tag{4.30}$$

By the projection technique again, it can be proved that the Y in question exists and is unique. This Y is called the *best linear predictor* and denoted by $\tilde{X}(t, s)$. This is the so called *linear prediction theory.* The prediction error in this case is given by

$$\sigma_l(t, s)^2 = E\{|X(t) - \tilde{X}(t, s)|^2\}. \tag{4.31}$$

In general, we have

$$\sigma_l(t, s)^2 \geq \sigma(t, s)^2, \qquad t, s \in I, \tag{4.32}$$

as is easily seen.

Linear prediction theory can be developed in the same manner as that for Gaussian processes, since only linear operations acting on the $X(t)$ are involved. More precisely, if X is a process satisfying the condition (4.26), then we can form a Gaussian process $Z = \{Z(t);\ t \in I\}$ with mean 0 and its covariance function $\Gamma(t, s)$ is equal to $E\{X(t)X(s)\}$. Consider a case where the process Z has a canonical representation. Then we can form the best predictor $\tilde{Y}(t, s)$ which is to be expressed in the form

$$\tilde{Z}(t, s) = f(Z(u);\ u \leq s, t),$$

with the help of the canonical representation. Note that f is linear in the $Y(u)$. The best linear predictor $\tilde{X}(t, s)$ for $X(t)$ based on the $X(u)$, $u \leq s$, has to be of the form

$$\tilde{X}(t, s) = f(X(u);\ u \leq s, t). \tag{4.33}$$

Alternative formulas are obtained as follows. Let $Z(t) = \int^t F(t, u)\, dB(u)$ be the canonical representation of $Z(t)$. Then, the best predictor $\tilde{Z}(t, s)$ is

$$\tilde{Z}(t, s) = \int^s F(t, u)\, dB(u).$$

Noting that there are involved only linear operations on the $X(t)$, we can play the same game as in (4.34) to obtain the expression for the best linear predictor of $X(t)$:

$$\tilde{X}(t, s) = \int^s F(t, u)\, d\xi(u), \tag{4.34}$$

where $d\xi(u)$ is a random measure, obtained from X, such that

$$E\{|dB(u)|^2\} = E\{|d\xi(u)|^2\}.$$

There are two particular classes of Gaussian processes for which canonical representations are explicitly formed; namely, the class of stationary Gaussian processes and that of multiple Markov Gaussian processes in the restricted sense (will appear in the next chapter). We now restrict our attention to the class of stationary processes which are not necessary Gaussian and apply what we have done using the results for the Gaussian case.

Let $X = \{X(t);\ t \in R^1\}$ be a stationary process satisfying the assumptions (A.1), (A.2′) and (A.3) in §§3.3–3.4. If X were a Gaussian process, then its canonical representation could be formed explicitly by Theorem 3.4. Our rephrasing technique gives us

$$\widetilde{X}(t, s) = \int_{-\infty}^{s} G(t-u)\, d\xi(u) \quad \text{with } E(|d\xi(u)|^2) = du. \tag{4.35}$$

The linear prediction error turns out to be

$$\sigma(t, s)^2 = \int_0^{t-s} G(u)^2\, du.$$

It is easier to find concrete formulas for predictors in the discrete parameter case. If a weakly stationary process $X = \{X_n;\ n \in I\}$ satisfies the assumptions (A.1) and (A.2), then we can easily see the counterpart of (4.35), by taking Theorem 3.3 into account. The details are omitted, but we mention one interesting formula here. If we consider the predictor for one unit of time in the future, then the prediction error is given by (3.30) in §3.3, and it is equal to the value $|c(0)|^2$. Hence, it is expressible as $|\sqrt{2\pi}\exp[g(0)]|^2$, which is eventually in agreement with

$$2\pi \exp\left[\frac{1}{2\pi}\int_{-\pi}^{\pi} |\log f(\lambda)|\, d\lambda\right].$$

Our final remark in this chapter is concerned with the case of higher multiplicity. Consider a Gaussian process with canonical representation of the form

$$X(t) = \sum_{i=1}^{N} \int^{t} F_i(t, u)\, dB_i(u)$$

without discrete spectrum. Then the best predictor $\widetilde{X}(t, s)$, $t > s$, is given by

$$\widetilde{X}(t, s) = \sum_{i=1}^{N} \int^{s} F_i(t, u)\, dB_i(u),$$

as a generalization of Proposition 4.2, and the prediction error therefore has the form

$$\sigma(t, s)^2 = \sum_{i=1}^{N} \int_{s}^{t} F_i(t, u)^2\, m_i(du).$$

CHAPTER V

Multiple Markov Gaussian Processes

In the present chapter we shall investigate the nature of dependency of distributions of Gaussian processes when the time parameter varies. The simplest case of such a dependency is observed in a sequence of independent random variables in the discrete time parameter case and by a white noise in the continuous time parameter case, although it is not an ordinary process but a generalized process with independent values at every point. On the other hand, there are Gaussian processes that heavily depend on the past in the sense that they are subject to strong dependency as time goes by. Such a dependency provides an essential viewpoint for observing stochastic processes. Thus, the property of dependency is specifically important in the investigation of Gaussian processes as well as in its applications.

The Markov property presents the most important property that describes the relation between past values and future values of a stochastic process. The class of Markov processes occupies a central position among all stochastic process, and the class is sitting just next to sequences of independent variables from the viewpoint of dependency.

Gaussian Markov processes with discrete parameter (or Markov chains) already have been discussed in Chapter 2. Intuitively speaking, a Markov process may be considered as a process that given its values before an instant t, then the conditional probability distribution of the future value depends only on the value of the process at the instant t. In other words, if we know the value of the Markov process at instant t, then the random phenomena before and after t are conditionally independent.

Then one may ask, when dependency is considered, what is a typical process close to a Markov process. A reasonable answer to this question is a multiple Markov process if we restrict our attention only to Gaussian processes. It is a process such that if its values are given in a neighborhood of an instant t, then its past and future become conditionally independent. Although the recipe is different, we can form canonical representations in both continuous and discrete parameter cases. Observations about the kernels of the representations lead us to introduce a class of weakly multiple Markov Gaussian processes. There one can find profound relations between analytic properties of the kernel and probabilistic properties of Gaussian processes.

With a view similar to that given by listing the Markov properties we can introduce the notion of *T-positivity* (or *reflection positivity*) of a Gaussian process. This notion arose originally from quantum field theory. It presents however, interesting probabilistic questions when considered from our present view-point.

It is worth mentioning that what we have proposed and discussed, multiple Markov properties and T-positivity, are only for Gaussian processes. We note that even though various attempts have been made to define multiple Markov properties of general stochastic processes, we still do not see much in the way of results.

§1. Multiple Markov processes with discrete parameter

First we define an N-ple Markov process, following Doob's idea, for general stochastic processes.

The parameter space I is taken to be the set of integers.

DEFINITION 5.1. A stochastic process $X = \{X_n(\omega);\ n \in I\}$ is said to be *at most N-ple Markov* if the conditional expectations satisfy

$$P(X_n \le x | \mathbf{B}_{n-1}(X)) = P(X_n \le x | X_{n-1}, X_{n-2}, \dots, X_{n-N}) \quad \text{a.e. } P, \qquad (5.1)$$

for every $n \in I$, where $\mathbf{B}_{n-1}(X) = \sigma\{X_i;\ i \le n-1\}$. The process X is called an *N-ple Markov process* if X is at most N-ple Markov process, but not at most $(N-1)$-ple Markov process.

We note that in Doob (1953), X is called an N-ple Markov process if it satisfies condition (5.1) only. We have actually introduced a somewhat restricted definition, which requires that the order of the Markov property is uniform in n, namely, N in (5.1) does not depend on n. With such a definition we shall be able to establish formulae for the canonical representation. As a slightly different definition, one might think of the following: If for any fixed n, the conditional probability $P(X_{n+k} \le x | \mathbf{B}_{n-1}(X))$, $k > 0$, is equal to the conditional probability $P(X_{n+k} \le x | X_{j_1}, X_{j_2}, \dots, X_{j_{N-1}}, X_{n-1})$, for any $j_1, j_2, \dots, j_{N-1} < n-1$, then $X = \{X_n\}$ is called the *N-ple Markov Gaussian process.* There $j_1, j_2, \dots, j_{N-1}$ can be picked arbitrarily, so that this definition seems to be very restricted. Indeed, a process X satisfying the above requirements is a very particular process, and hence the definition just proposed can never be an essential extension of a simple Markov process. This fact is easily seen by using the canonical representation of X. With such an observation we can see that the above definition is a plausible extension of the simple Markov property for a Gaussian process.

If X is an at most N-ple Markov Gaussian process, then the vector valued process $\mathbf{X} = \{\mathbf{X}_n = (X_n, X_{n-1}, \dots, X_{n-N+1});\ n \in I\}$ is a vector-valued Markov process in the ordinary sense. As an additional remark, we claim that (5.1) is equivalent to

$$P(X_{n+k} \le x|\mathbf{B}_{n-1}(X)) = P(X_{n+k} \le x|X_{n-1}, X_{n-2}, \dots, X_{n-N}), \quad \text{a.e. } P, \quad k = 0, 1, 2, \dots . \tag{5.2}$$

Proposition 5.1. *If an N-ple Markov Gaussian process X satisfies assumptions* (A.1) *and* (A.2) *in Chapter* III, *then there exists a canonical representation whose kernel is expressed in the form*

$$\sum_{i=1}^{N} a_n^i b_j^i, \qquad j \le n. \tag{5.3}$$

Moreover, the kernel can never be expressed as a sum, such as (5.3), *of less than N terms.*

Remark. This is a generalization of Theorem 2.6.

Proof. The existence of a canonical representation is known from the fact that $H_n(X) - H_{n-1}(X)$ is at most one-dimensional (actually just one dimensional). Thus we can directly construct the innovation process $\{\xi_n ; n \in I\}$, and the canonical representation follows.

The expression (5.3) for the canonical kernel can be derived from condition (5.2). The details of the proof are omitted. □

We note that the proof of Proposition 5.1. is analogous, as will be seen later, to that of Theorem 5.1 which gives properties of continuous parameter processes.

We now give a typical example of an N-ple Markov Gaussian process. Let $X = \{X_n ;\ n \in I\}$ be a complex stationary Gaussian process satisfying assumptions (A.1) and (A.2) in Chapter III. Then the density function $f(\lambda)$ of the spectral distribution exists and is expressed in the form

$$f(\lambda) = \frac{1}{2\pi}\left\{\left|\sum_{j=0}^{N} b_i e^{-ij\lambda}\right|\right\}^2, \qquad b_0 \ne 0, \tag{5.4}$$

with a suitable choice of a polynomial $P(z) = \sum_{j=0}^{N} b_j z^j$. Since $f(\lambda)$ is integrable over $[0, 2\pi]$, the polynomial $P(z)$ has no zeroes on the unit circle $|z| = 1$. The roots of $P(z) = 0$ are therefore either inside or outside of the circle. In fact, the discussion in §3.3, claims that to obtain the canonical kernel all of roots should be outside of the unit circle. If there were a point z_0 such that $P(z_0) = 0$ with $|z_0| < 1$, then we would replace $P(z) = (z - z_0)P_1(z)$ by $P(z) = (z\overline{z}_0 - 1)P_1(z)$, which has the root $1/\overline{z}_0$ sitting outside of the circle and has the same modulus as $P(z)$ on the circle. We may therefore assume that $P(z)$ has not roots inside of the unit circle.

On the other hand, X_n has the spectral representation

$$X_n = \int_{-\pi}^{\pi} e^{in\lambda}\, dZ(\lambda), \qquad E|dZ(\lambda)|^2 = f(\lambda)\, d\lambda, \tag{5.5}$$

where dZ is a complex Gaussian random measure. By using the coefficients $b_0, b_1, \ldots, b_N$ of $P(z)$, we can define the innovation $\{\xi_n;\ n \in I\}$ in such a way that

$$\xi_n = \sum_{j=0}^{N} b_j X_{n-j} \qquad \left(= \int_{-\pi}^{\pi} P(e^{-i\lambda}) e^{in\lambda}\, dZ(\lambda) \right). \tag{5.6}$$

The $\{\xi_n\}$ is a system of independent standard complex Gaussian random variables and each ξ_n belongs to $H_n(X)$. These properties can be proved in the following manner. For any $k \geq 0$ we have

$$\begin{aligned} E(X_{n-k}\overline{\xi}_{n+1}) &= \int_{-\pi}^{\pi} e^{i(n-k)\lambda} \sum_{j=0}^{N} \overline{b}_j e^{-i(n+1-j)\lambda} f(\lambda)\, d\lambda \\ &= \int_{-\pi}^{\pi} e^{-ik\lambda} \frac{e^{-i\lambda}}{P(e^{-i\lambda})}\, d\lambda = -\int_{|z|=1} z^k \frac{1}{P(z)}\, dz = 0, \end{aligned}$$

since $P(z)$ has no roots inside of the unit disk. This proves that ξ_{n+1} is orthogonal to $H_n(X)$. It is easy to see that

$$E\{|\xi_{n+1}|^2\} = \int_{-\pi}^{\pi} |P(e^{i\lambda})|^2 f(\lambda)\, d\lambda = 1.$$

While, we obtain

$$X_{n+1} = -b_0^{-1}(b_1 X_n + \cdots + b_N X_{n-N+1}) + b_0^{-1}\xi_{n+1},$$

which means that X_{n+1} is expressed as a sum of an element of $H_n(X)$ and $b_0^{-1}\xi_{n+1}$. Thus, $\{\xi_n;\ n \in I\}$ has been proved to be an innovation process for X.

We are now ready to form the canonical kernel by a concrete method. For convenience, we assume that $P(z) = 0$ has no multiple roots in order to have the following expression

$$P(z) = b_0(z - \alpha_1)(z - \alpha_2)\cdots(z - \alpha_N), \qquad |\alpha_j| > 1, \quad j = 1, 2, \ldots, N, \tag{5.7}$$

where the α_j are different.

Introduce difference operators Δ_j by $\Delta_j x_n = -\alpha_j x_n + x_{n-1}$, $j = 1, 2, \ldots, N$, acting on the space of sequences $\{x_n\}$. Using these operators, (5.6) may be written as a stochastic difference equation of the following form:

$$b_0 \Delta_N \Delta_{N-1} \cdots \Delta_1 X_n = \xi_n, \qquad n \in I.$$

The inverse operator of Δ_N exists, and we are given

$$b_0 \Delta_{N-1} \cdots \Delta_1 X_n = \sum_{j=-\infty}^{n} \alpha_N^{-n+j-1} \xi_j.$$

We can play the same game to obtain

$$\begin{aligned} b_0\Delta_{N-2}\cdots\Delta_1 X_n &= \sum_{k=-\infty}^{n} \alpha_{N-1}^{-n+k-1} \sum_{j=-\infty}^{k} \alpha_N^{-k+j-1}\xi_j \\ &= \sum_{j=-\infty}^{n} \left\{ \frac{\alpha_{N-1}^{-(n-j)}}{\alpha_{N-1}(\alpha_N - \alpha_{N-1})} - \frac{\alpha_N^{-(n-j)}}{\alpha_N(\alpha_N - \alpha_{N-1})} \right\} \xi_n . \end{aligned}$$

Repeating such a calculation, we finally come to

$$X_n = \sum_{j=-\infty}^{n} \{c_1\alpha_1^{-(n-j)} + c_2\alpha_2^{-(n-j)} + \cdots + c_N\alpha_N^{-(n-j)}\}\xi_j, \quad \text{with } \sum_{j=1}^{N} c_i = b_0^{-1}. \tag{5.8}$$

It is easy to show that (5.8) gives the canonical representation and the kernel is of the form (5.3) and $a_n^i = c_i\alpha_i^{-n}$ and with $b_j^i = \alpha_i^j$.

Therefore, X satisfies (5.1) and is an N-ple Markov Gaussian process.

In the case above, the prediction error per one step, which is $\sigma_1^2 = E((X_{n+1} \hat{X}(n, 1)^2)$, is equal to $|b_0|^{-2}$.

The process discussed above is a most typical and elementary example of a process having the N-ple Markov property. Historically speaking, it was G. U. Yule (1927) who observed the Wolf number (sun spot number) of the sun and found that the stochastic process that arises from the sun spot number is a stationary process. Moreover, he tried to describe such random phenomenon as the maculae numerically by a mathematical model. The model is represented in terms of a stationary process X_n satisfying a stochastic difference equation of the form

$$X_n + a_1X_{n-1} + a_2X_{n-2} = \xi_n$$

with a suitable choice of coefficients a_1 and a_2, where $\{\xi_n\}$ is an independent random sequence. This attempt by Yule brought about great progress in the investigation of stationary processes, and it enables us now to recognize that a stochastic process satisfying a stochastic difference equation is important, not only in applied fields, but also in the theory of probability (H. Wold (1938)). Such a historical note emphasizes the point that the multiple Markov Gaussian processes illustrated in this section occupy an important class.

§2. Multiple Markov Gaussian processes with continuous parameter

Before we come to the topic in the title of this section we consider the notion of the multiple Markov property in the case of continuous parameter as compared to a discrete parameter.

For a Gaussian process $X = \{X(t, \omega);\ t \in I\}$ to be N-ple Markov process, we expect, as in §1, that an N-dimensional vector-valued simple Markov process $\hat{X} = \{\hat{X}(t);\ t \in I\}$ can be formed from the given process X. If, in particular, X is $N-1$ times differentiable in some sense of convergence and if $\{\widehat{X}(t) = (X(t), X'(t), \ldots, X^{(N-1)}(t));\ t \in I\}$ is a simple Markov process, then we would call X an N-ple Markov Gaussian process. This would be an easy way to define a multiple Markov property. Indeed, J. L. Doob (1944) introduced an N-ple Markov property for stationary Gaussian processes from the above standpoint. His definition is as follows. Let $X(t)$ be a real stationary Gaussian process. If $X(t)$ is $(N-1)$-times differentiable in the mean square sense and if X satisfies a differential equation

$$\sum_{k=0}^{N} a_k \frac{d^{N-k}}{dt^{N-k}} X(t) = \dot{B}_0(t), \qquad \dot{B}_0(t) = \frac{d}{dt} B_0(t), \tag{5.9}$$

then X is called an N-ple *Markov Gaussian process*. Here B_0 is a Brownian motion. The N-th derivative of $X(t)$ and derivative of $B_0(t)$ are taken in the sense of distribution.

Equation (5.9) is regarded as a natural extension of the *Langevin equation*

$$a_0 X'(t) + a_1 X(t) = \dot{B}_0(t), \qquad a_0 \cdot a_1 > 0,$$

which is satisfied by a stationary simple Markov Gaussian process.

On the other hand, P. Lévy (1956) called X an N-ple *Markov Gaussian process in the restricted sense* if X is $N-1$ times differentiable and if the conditional expectation $E(X(t)|\mathbf{B}_s(X))$, $s < t$, is a function of only $X(s)$, $X's)$, $\ldots$, $X^{(N-1)}(s)$, where $\mathbf{B}_s(X) = \sigma\{X(u);\ u \le s\}$. Note that $E(X(t)|\mathbf{B}_s(X))$ is a linear combination of $X(s)$, $X'(s)$, $\ldots$, $X^{(N-1)}(s)$, since X is Gaussian. This definition is in agreement with Doob's definition if X is a stationary Gaussian process.

These definitions describe the multiple Markov property for Gaussian processes from somewhat different standpoints, both depending on the linear structure of X as a Gaussian system. In each case, however, we may be discontent due to the unwelcome assumption on the differentiability of $X(t)$, which does not seem to be directly connected with the manner of dependency of X as the time t varies.

While we will see a notable characteristic as soon as the kernel of the canonical representation of an N-ple Markov Gaussian process is obtained. In fact, the kernel $F(t, u)$ of the canonical representation is a continuous analogue of (5.3) in Proposition 5.1. of the preceding section; in other words, $F(t, u)$ is a *Goursat kernel* of degree N:

$$F(t, u) = \sum_{i=1}^{N} f_i(t) g_i(u), \qquad u \le t. \tag{5.10}$$

With such a kernel, the canonical representation tells us that N additive processes are causally constructed from X, and X itself is expressed as a linear combination of the N additive processes

$$X(t) = \sum_{i=1}^{N} f_i(t) U_i(t) \quad \text{with } U_i(t) = \int^t g_i(u)\, dB_0(u).$$

As for the conditional expectation, we have

$$E(X(t)|\mathbf{B}_s(X)) = \sum_{i=1}^{N} f_i(t)U_i(s), \qquad t \geq s,$$

which is again a linear combination of $U_i(s)$, $1 \leq i \leq N$, for any t. So far as processes are restricted to be Gaussian, that is in our case, the above expression seems to possess the characteristic, which we are in search of, for defining the multiple Markov property and it is even free from differentiability. Thus, we come to the following definition.

DEFINITION 5.2. A Gaussian process $X = \{X(t, \omega);\ t \in I\}$ is called an *N-ple Markov Gaussian process*, if $E(X(t_i)|\mathbf{B}_{t_0}(X))$, $i = 1, 2, \ldots, N$, are linearly independent in $H(X)$ for any $t_0 \leq t_1 < t_2 < \cdots < t_N$ and if $E(X(t_i)|\mathbf{B}_{t_0}(X))$, $i = 1, 2, \ldots, N+1$, are linearly dependent for any $t_0 \leq t_1 < t_2 < \cdots < t_{N+1}$.

A simple Markov Gaussian process is a 1-ple Markov process provided that the covariance never vanishes, which is seen from the discussion in §2.5.

REMARK. The definition of an N-ple Markov Gaussian process is close to that of an N-ple Markov process in the weak sense of P. Lévy. In fact, our definition is slightly stronger than Lévy's, because the Markov property, defined in Definition 5.1, requires uniformity in time. For example, if X is a simple Markov process, then $X(t)$ may be independent of some $X(s)$, $s \leq t_0$, but such a situation is not allowed for a 1-ple Markov Gaussian process.

EXAMPLE 1. A Gaussian process $X = \{X(t, \omega);\ t \in I\}$ given by

$$X(t) = \int_0^t (2t - u)\, dB_0(u)$$

is a double Markov process. In fact, $2t - u$ is a canonical kernel and for any $t_0 \leq t_1 < t_2 < t_3$,

$$E(X(t_i)|\mathbf{B}_{t_0}(X)) = 2t_i B_0(t_0) - \int_0^{t_0} u\, dB_0(u), \qquad i = 1, 2, 3.$$

Since these conditional expectations are linear combinations of $B_0(t_0)$ and $\int_0^{t_0} u\, dB_0(u)$, the three are linearly dependent in $H(X)$, and we can choose any pair to be a linearly independent system. This means that X is a double Markov process. However, $X'(t)$ does not exist so it is not a double Markov in the restricted sense.

The next step is concerned with the representation of an N-ple Markov Gaussian process. The conditions (A.1), (A.2′), and (A.3) are assumed. In addition, we assume the following condition:

$$H_t(X) \text{ is continuous in } t. \tag{A.4}$$

This means that $H_t(X) = \bigvee_{s<t} H_s(X) = \bigcap_{s>t} H_s(X)$, $t \in I$.

Note that the condition (A.4) implies that X does not have any discrete spectrum in the expression of the canonical representation.

THEOREM 5.1. *Let a Gaussian process* $X = \{X(t, \omega);\ t \in I\}$ *have a canonical representation* $X(t) = \int^t F(t, u)\,dB(u)$ *of unit multiplicity. Then,* X *is an* N*-ple Markov process if and only if the conditions* (i) *and* (ii) *are satisfied.*

(i) *the kernel* $F(t, u)$ *is expressed in the form* (5.10) *satisfying*

$$\det(f_i(t_j)) \neq 0, \qquad t_j \in I^0 = \text{ the interior of } I, \tag{5.11}$$

for any N*-tuple of different* $t_1, t_2, \ldots, t_N$, *and*

(ii) $g_i(u)$, $i = 1, 2, \ldots, N$, *are linearly independent in the Hilbert space*

$$L^2(m, t) = \left\{f;\ \int_{u \le t} f(u)^2\, m(du) < \infty\right\}$$

for any $t \in I$, *where* $m(dt) = E(dB(t)^2)$.

PROOF. Assume that X has canonical representation $X(t) = \int^t F(t, u)dB(u)$, namely $H_t(X) = H_t(B)$ holds for every $t \in I$.

1. *Proof of necessity.* Suppose that X is an N-ple Markov Gaussian process. By definition, for any $t_0 = t_1 < t_2 < \cdots < t_N < \tau$, there exist $a_j(\tau, \mathbf{t}) = (\mathbf{t} = (t_1, t_2, \ldots, t_N))$, $j = 1, 2, \ldots, N$, such that

$$E(X(\tau)|\mathbf{B}_{t_1}(X)) = \sum_{j=1}^{N} a_i(\tau, \mathbf{t}) E(X(t_j)|\mathbf{B}_{t_1}(X)) = 0.$$

Since the representation is canonical, we obtain

$$\int^{t_1} \left\{F(\tau, u) - \sum_{j=1}^{N} a_j(\tau, \mathbf{t}) F(t_j, u)\right\} dB(u) = 0.$$

Therefore, we know that for any $\tau > t_N$,

$$F(\tau, u) - \sum_{j=1}^{N} a_j(\tau, \mathbf{t}) F(t_j, u) = 0 \tag{5.12}$$

holds almost everywhere in $u \le t_1$.

Note that the above equality is valid almost everywhere in u with respect to the m-measure. Such remarks will be omitted in what follows.

Next taking $\mathbf{s} = (s_1, s_2, \ldots, s_N)$, $s_1 < s_2 < \cdots < s_N < t_1$, instead of $\mathbf{t}$ and taking $X(t_j)$ in place of $X(\tau)$, we can play the same game with (5.12) to obtain

$$F(t_j, u) - \sum_{k=1}^{N} a_k(t_j, \mathbf{s}) F(s_k, u) = 0, \qquad u \le s_1, \quad j = 1, 2, \ldots, N,$$

$$F(\tau, u) - \sum_{k=1}^{N} a_k(\tau, \mathbf{s}) F(s_k, u) = 0, \qquad u \le s_1. \tag{5.13}$$

Combining (5.13) with (5.12),

$$\sum_{k=1}^{N} a_k(\tau, \mathbf{s})F(s_k, u) = \sum_{j=1}^{N}\sum_{k=1}^{N} a_j(\tau, \mathbf{t})a_k(t_j, \mathbf{s})F(s_k, u), \qquad u \leq s,$$

is obtained. From Definition 5.2, $E(X(s_k)|\mathbf{B}_{s_1}(X))$, $k = 1, 2, \dots, N$, are linearly independent in $H(X)$, so are $F(s_k, u)$, $k = 1, 2, \dots, N$, in $L^2(m, s_1)$. Hence, for any k

$$a_k(\tau, \mathbf{s}) = \sum_{i=1}^{N} a_i(\tau, \mathbf{t})a_k(t_i, \mathbf{s}). \tag{5.14}$$

Furthermore, since the $F(t_i, u)$ are linearly independent, $\det(a_k(t_i, \mathbf{s})) \neq 0$ is derived from (5.13). Hence, there exists a square matrix $B(s, \mathbf{t})$ of degree N satisfying

$$\mathbf{a}(\tau, \mathbf{t}) = a(\tau, \mathbf{s})B(\mathbf{s}, \mathbf{t}), \qquad B(\mathbf{s}', \mathbf{t}) = B(\mathbf{s}'), \mathbf{s})B(\mathbf{s}, \mathbf{t}), \tag{5.15}$$

where $a(\tau, \mathbf{t}) = (a_1(\tau, \mathbf{t}), a_2(\tau, \mathbf{t}), \dots, a_N(\tau, \mathbf{t}))$ and $\mathbf{s}' = (s_1', s_2', \dots, s_N')$, with $s_1' < s_2' < \cdots < s_N' < s_1$. Now fix $t_1, t_2, \dots, t_N$ and define a vector $\mathbf{f}_s = (f_{s_1}, f_{s_2}, \dots, f_{s_N})$ in such a way that

$$\mathbf{f}_\mathbf{s}(\tau) = \mathbf{a}(\tau, \mathbf{s})B(\mathbf{s}, \mathbf{t}), \qquad \tau \geq s_N.$$

Then we see that $\mathbf{f}_{\mathbf{s}'}(\tau)$ is an extension of $\mathbf{f}_\mathbf{s}(\tau)$ as a function of τ. Thus, a vector-valued function $\mathbf{f}(\tau) = (f_1(\tau), f_2(\tau), \dots, f_N(\tau))$, $t \in I^0$, is uniquely determined. It is easy to see that the $f_i(t)$, $i = 1, 2, \dots, N$, satisfy condition (5.11). If $\tau > s_N$ and $u < s_1$, then we have

$$F(\tau, u) = \mathbf{a}(\tau, \mathbf{t})\mathbf{F}(\mathbf{t}, u)^* = \mathbf{f}(\tau)B(\mathbf{s}, \mathbf{t})^{-1}\mathbf{F}(\mathbf{s}, u)^* = \mathbf{f}(\tau)\mathbf{g}_\mathbf{s}(u, t)^*,$$

where * denotes the transposed vector,

$$\mathbf{F}(\mathbf{t}, u) = (F(t_1, u), F(t_2, u), \dots, F(t_N, u)),$$

and

$$\mathbf{g}_\mathbf{s}(u, \tau) = \mathbf{F}(\mathbf{s}, u)B(\mathbf{s}, \mathbf{t})^{-1}.$$

This $\mathbf{g}_\mathbf{s}(u, \mathbf{t})$ has so far been defined only for $u < s_1$, but it can be extended to a vector-valued function $\mathbf{g}(u) = (g_1(u), g_2(u), \dots, g_N(u))$ defined on I^0, by the analogous method to the determination of $\mathbf{f}(\tau)$ from $\mathbf{f}_\mathbf{s}(\tau)$. Thus, $F(t, u)$, $u < t$, is expressed in the form (5.10). Note that we have the same expression (5.10) even for $u \leq t$, because of the assumption (A.4).

2. *Proof of sufficiency.* Suppose a canonical kernel $F(t, u)$ is given by (5.10). For any $t_0 \leq t_1 < t_2 < \cdots < t_N < t_{N+1}$, the $E(X(t_i)|\mathbf{B}_{t_0}(X))$ $i = 1, 2, \dots, N+1$, are all linear combinations of

$$\int^{t_0} g_i(u)\,dB(u), \qquad i = 1, 2, \dots, N,$$

in $H_{t_0}(X)$, and they form a system of linearly independent Gaussian random variables. The latter condition of Definition 5.2 is therefore satisfied. The former condition is trivially satisfied because of (5.11) and of the property that the $g_i(u)$ are linearly independent in $L^2(m, t)$ for any $t \in I$. Thus, we have proved that X is an N-ple Markov Gaussian process. □

If $\mathbf{f} = \{f_i(t)\}$ is a system of functions satisfying condition (5.11), then $\mathbf{f}$ is said to satisfy the *total positivity* property.

COROLLARY. *Let X be an N-ple Markov Gaussian process, and let the canonical kernel $F(t, u)$ be given by* (5.10). *Then the covariance function $\Gamma(t, s)$ has the form*

$$\Gamma(t, s) = \sum_{i=1}^{N} f_i(t) h_j(s), \qquad s \le t, \tag{5.16}$$

where f_i, $i = 1, 2, \ldots, N$, satisfying (5.11) *and $(h_1(s), h_2(s), \ldots, h_N(s))$ is expressible as $(f_1(s), f_2(s), \ldots, f_N(s))G(s)$ with $G(s) = (g_{ij}(s))$ some Gramian matrix.*

Furthermore, $\Gamma(t, s)$ cannot be expressed as a sum of less than N terms in the form (5.16).

PROOF. Theorem 5.1, together with the canonical kernel (5.10), yields

$$\Gamma(t, s) = E(X(t)X(s)) = \sum_{i=1}^{N} f_i(t) \left\{ \sum_{j=1}^{N} f_j(s) \int^{s} g_i(u) g_j(u)\, m(du) \right\},$$

for $t > s$. Set $g_{ij}(s) = \int^s g_i(u) g_j(u)\, m(du)$ for $i, j = 1, 2, \ldots, N$, to form a Gramian matrix $G(s) = (g_{ij}(s))$. Then, we obtain (5.16).

To prove the last assertion it is sufficient to show that $h_i(s)$, $i = 1, 2, \ldots, N$ are linearly independent. Suppose that they are linearly dependent. Then, there exist constants $c_1, c_2, \ldots, c_N$ such that $\sum_{i=1}^N c_i h_i(s) \equiv 0$, or equivalently $\int^s (\sum_{i=1}^N c_i g_i(u)) F(s, u) m(du) \equiv 0$. Since $F(s, u)$ is a canonical kernel, we obtain $\sum_{i=1}^N c_i g_i(u) \equiv 0$ from Theorem 4.4. This contradicts the condition on $g_1, g_2, \ldots, g_N$ in Theorem 5.1. □

As we have observed, the N-ple Markov property can be described in terms of the canonical kernel. Thus we have the following definition.

DEFINITION 5.2. If a kernel $F(t, u)$ is expressed in the form of (5.10) using $f_i(t)$, $i = 1, 2, \ldots, N$, satisfying the total positivity condition, and $g_i(u)$, $i = 1, 2, \ldots, N$, which are linearly independent in $L^2(m; t)$ for any $t \in I$, then $F(t, u)$ is called a *Goursat kernel of order N* (*with respect to* m).

With this definition, Theorem 5.1 is rephrased as follows: *A Gaussian process having a canonical representation has the n-ple Markov property if and only if the canonical kernel is a Goursat kernel of order N.*

We note that a Goursat kernel is not always a canonical kernel, for instance, the kernel in Example B(ii) in §4.4 is a Goursat kernel but not a canonical one.

Except for the case $N = 1$, the N-ple Markov property does not guarantee the existence of a canonical representation (with unit multiplicity). However, under the assumptions (A.1) and (A.2$'$), the multiplicity of an N-ple Markov process is at most N. L. Pitt (1975) and V. Mandrekar (1974) present more detailed results on these topics.

Here is a simple example.

EXAMPLE 2. Let $\{B_1(t, \omega);\ t \geq 0\}$ and $\{B_2(t, \omega);\ t \geq 0\}$ be mutually independent Brownian motions, and let $F(t)$ be a monotonically increasing and absolutely continuous function on $[0, \infty)$. Assume that the Radon-Nikodym derivative $F'(t)$ is not square integrable on any nonempty open interval. Then, a Gaussian process $X = \{X(t, \omega);\ t \geq 0\}$ defined by

$$X(t) = B_1(t) + F(t)B_2(t) \tag{5.17}$$

is a particular example of processes introduced in part D of §4.4, and it has double multiplicity. The canonical representation is clearly given by (5.17). For $0 < t_0 \leq t_1 < t_2 < t_3$, we have

$$E(X(t_i)|\mathbf{B}_{t_0}(X)) = B_1(t_0) + F(t_i)B_2(t_0), \qquad t = 1, 2, 3.$$

The three conditional expectations are linearly dependent, but any two of them are linearly independent. This means that X is a *double* Markov Gaussian process.

§3. Multiple Markov Gaussian processes in the restricted sense

An N-ple Markov Gaussian process which is $(N-1)$-times differentiable in the mean square sense is essentially an N-ple Markov process in a restricted sense defined by P. Lévy. The significance of such a process is that the canonical representation always exists and its innovation process can be formed by using the well-known technique of the theory of ordinary differential equations. Moreover, Green's function is taken to be the canonical kernel, so that the structure of the given process can be investigated in an analytic manner.

Throughout this section, the parameter space I is taken to be the half line $[0, \infty)$. The stationary case is therefore excluded, but change of time scale converts stationary processes with the parameter space $(-\infty, \infty)$ to those processes with the parameter space $[0, \infty)$ without changing the Markov property, so that the above assumption for I is not essential.

Before introducing an N-ple Markov Gaussian process in the restricted sense, we review some of the properties of ordinary differential operators (see, e.g., E. L. Ince (1926)).

Take real-valued functions $v_i(t)$, $i = 0, 1, 2, \ldots, N$, defined on I satisfying the conditions

$$v_i(t) \in \mathbf{C}(I), \qquad i = 0, 1, 2, \ldots, N, \tag{5.18}$$

$$v_i(t) \neq 0, \qquad t > 0, \qquad i = 0, 1, 2, \ldots, N. \tag{5.19}$$

We introduce the following various (in somewhat generalized sense) differential operators

$$L_t = \frac{1}{v_0(t)}\frac{d}{dt}\frac{1}{v_1(t)}\frac{d}{dt}\cdots\frac{1}{v_{N-1}(t)}\frac{d}{dt}\frac{1}{v_N(t)},$$
$$L_t^{(j)} = \frac{1}{v_j(t)}\frac{d}{dt}\frac{1}{v_{j+1}(t)}\frac{d}{dt}\cdots\frac{1}{v_{N-1}(t)}\frac{d}{dt}\frac{1}{v_N(t)}, \qquad j = 1, 2, \dots, N-1, \tag{5.20}$$
$$L_u^* = \frac{1}{v_N(u)}\frac{d}{du}\frac{1}{v_{N-1}(u)}\frac{d}{du}\cdots\frac{1}{v_1(u)}\frac{d}{du}\frac{1}{v_0(u)}.$$

If each v_i is sufficiently smooth, then L_t, $L_t^{(j)}$, and L_u^* are all differential operators in the usual sense. Note that L_u^* is the (formal) adjoint operator for L_t. Needless to say, the expression of L_t in terms of $\{v_i;\ i = 1, 2, \dots, N\}$ is not unique; indeed there is much freedom in the choice of $\{v_i\}$ for a given operator L_t. For example, $\{v_i\}$ may be chosen in the following manner: having a fundamental system $\{f_i(t);\ i = 1, 2, \dots, N\}$ of solutions of a differential equation

$$L_t f(t) = 0, \tag{5.21}$$

we can find $\{v_i\}$ such that

$$f_1(t) = v_N(t),$$
$$f_i(t) = v_N(t)\int_0^t v_{N-1}(t_1)\,dt_1 \int_0^{t_1} v_{N-2}(t_2)\,dt_2 \int_0^{t_2} \cdots \int_0^{t_{i-2}} v_{N-i+1}(t_{i-1})\,dt_{i-1}, \qquad i = 2, \dots, N. \tag{5.22}$$

Conversely, for a given system $\{v_i(t)\}$ of $\mathbf{C}^\infty(T)$-functions with property (5.19), we can define L_t and $\{f_i(t);\ i = 1, 2, \dots, N\}$ by (5.20) and (5.22), respectively, so that $\{f_i\}$ is a system of fundamental solutions for the differential equation (5.21).

Again, starting from the system $\{v_i;\ i = 1, 2, \dots, N\}$, we can define an operator L_u^* by using (5.20) where the g_i are given by

$$g_i(u) = (-1)^{N-i} v_0(u)\int_0^u v_1(u_1)\,du_1 \int_0^{u_1} v_2(u_2)\,du_2 \int_0^{u_2} \cdots \int_0^{u_{N-i-1}} v_{N-i}(u_{N-i})\,du_{N-i}, \qquad i = 1, 2, \dots, N-1, \tag{5.23}$$
$$g_N(u) = v_0(u).$$

$\{g_i(u)\}$ is a fundamental system of solutions of the equation

$$L_u^* g(u) = 0. \tag{5.24}$$

REMARK. If the lower limits of the integrals (5.22) and (5.23) defining the systems $\{f_i\}$ and $\{g_j\}$ are replaced by any other constants, the two systems still give fundamental systems of solutions for the respective differential equations.

By using these systems $\{f_i\}$ and $\{g_j\}$ in (5.22) and (5.23), we define a function $R(t, u)$ on $I \times I$ as follows:

$$R(t, u) = \begin{cases} \sum_{i=1}^{N} f_i(t) g_i(u), & u \le t, \\ 0, & u > t. \end{cases} \tag{5.25}$$

It is clear that $R(t, u)$ is continuous not only for $t \ne u$ but also on the diagonal. Moreover, we have

Lemma 5.1. *The function $R(t, u)$ has the following properties:*

$$\begin{aligned} L_t R(t, u) &\equiv 0, \quad t > u \text{ for any fixed } u, \\ L_u^* R(t, u) &\equiv 0, \quad u < t \text{ for any fixed } t. \end{aligned} \tag{5.26}$$

and

$$\begin{aligned} &R(t, t) = 0, \\ &L_t^{(j)} R(t, u)|_{u=t} \equiv 0, \qquad j = 2, 3, \dots, N-1, \\ &L_t^{(1)} R(t, u)|_{u=t} \ne 0. \end{aligned} \tag{5.27}$$

Proof. The equation (5.26) is easily proved, since f_i and g_i are solutions to (5.21) and (5.24), respectively. Next, from the construction of the f_i, it follows that

$$L_t^{(1)} f_N(t) \equiv 1,$$

and

$$L_t^{(1)} f_i(t) \equiv 0, \qquad i = 1, 2, \dots, N-1.$$

Noting that $g_N(t) = v_0(t) \ne 0$, the last equality of (5.27) is proved. The first two are easily proved by mathematical induction. □

Lemma 5.2. *If $\varphi(u)$ in $L^2(I)$ satisfies*

$$\int_0^t R(t, u)\varphi(u)du \equiv 0, \qquad t \in (a, b), \tag{5.28}$$

then $\varphi(u) = 0$, a.e. over the interval (a, b).

Proof. The assumption (5.28) implies

$$\sum_{i=1}^{N} f_i(t) \int_0^t g_i(u)\varphi(u)\, du \equiv 0, \qquad t \in (a, b).$$

Divide the left-hand side by $v_N(t)$ and obtain its absolutely continuous factor. Then taking the Radon-Nikodym derivative, we have

$$\sum_{i=1}^{N} f_i^{(1)}(t) \int_0^t g_i(u)\varphi(u)\, du + \left\{ \sum_{i=1}^{N} f_i(t) g(t) \right\} \varphi(t) \equiv 0, \qquad t \in (a, b).$$

Note that $f_N^{(1)}(t) \equiv 0$, where $f^{(1)}$ means

$$\frac{d}{dt}\frac{1}{v_N(t)}f(t).$$

The second term is identically equal to zero because of Lemma 5.1, and hence we obtain

$$\sum_{i=1}^{N} f_i^{(1)}(t)\int_0^t g_i(u)\varphi(u)\,du \equiv 0, \qquad t \in (a, b).$$

Further, dividing this formula by $v_{N-1}(t)$ and taking the derivative again, we are given

$$\sum_{i=1}^{N-2} f_i^{(2)}(t)\int_0^t g_i(u)\varphi(u)\,du \equiv 0, \qquad t \in (a, b).$$

Repeating this argument, we finally come to

$$\int_0^t g_N(u)\varphi(u)\,du = \int_0^t v_0(u)\varphi(u)\,du \equiv 0, \qquad t \in (a, b).$$

Since $v_0(t) \neq 0$, we have $\varphi(u) = 0$ a.e., which was to be proved. □

It is known that the kernel $R(t, u)$ is Green's function for L_t of Volterra type. For $\varphi \in L^2(I)$, the function

$$F(t) = \int_0^t R(t, u)\varphi(u)\,du$$

belongs to the domain of the operator L_t and satisfies

$$L_t F(t) = \varphi(t). \tag{5.29}$$

LEMMA 5.3. (i) *The system* $\{f_i(t);\ i = 1, 2, \dots, N\}$ *given by* (5.22) *enjoys the total positivity property in the sense that*

$$\det(f_i(s_j)) \neq 0, \tag{5.30}$$

for any different $s_1, s_2, \dots, s_N$.

(ii) *The system* $\{g_i(u);\ i = 1, 2, \dots, N\}$ *given by* (5.23) *consists of linearly independent members in* $L^2([0, t])$ *for every* $t > 0$.

PROOF. (i) Suppose that there were N different points $s_1 < s_2 < \cdots < s_N$ such that $\Delta = \det(f_i(s_j)) = 0$. Noting that

$$\prod_{i=1}^{N} v_i(s_j) \neq 0,$$

we observe the following equalities that come from the assumption:

$$\left\{\prod_{i=1}^{N} v_i(s_j)\right\}^{-1} \Delta$$

$$= \begin{vmatrix} 1 & 1 & \cdots & 1 \\ \int_0^{s_1} v_{N-1}\,dt & \int_0^{s_2} v_{N-1}\,dt & \cdots & \int_0^{s_N} v_{N-1}\,dt \\ \int_0^{s_1} v_{N-1}\int v_{N-2}\,(dt)^2 & \int_0^{s_2} v_{N-1}\int v_{N-2}\,(dt)^2 & \cdots & \int_0^{s_N} v_{N-1}\int v_{N-2}\,(dt)^2 \\ \cdots & \cdots & \cdots & \cdots \\ \int_0^{s_1} v_{N-1}\int\cdots\int v_1\,(dt)^{N-1} & \int_0^{s_2} v_{N-1}\int\cdots\int v_1\,(dt)^{N-1} & \cdots & \int_0^{s_N} v_{N-1}\int\cdots\int v_1\,(dt)^{N-1} \end{vmatrix}$$

$$= \begin{vmatrix} \int_{s_1}^{s_2} v_{N-1}\,dt & \int_{s_2}^{s_3} v_{N-1}\,dt & \cdots & \int_{s_{N-1}}^{s_N} v_{N-1}\,dt \\ \int_{s_1}^{s_2} v_{N-1}\int v_{N-2}\,(dt)^2 & \int_{s_2}^{s_3} v_{N-1}\int v_{N-2}\,(dt)^2 & \cdots & \int_{s_{N-1}}^{s_N} v_{N-2}\int v_{N-2}\,(dt)^2 \\ \cdots & \cdots & \cdots & \cdots \\ \int_{s_1}^{s_2} v_{N-1}\int\cdots\int v_1\,(dt)^{N-1} & \int_{s_2}^{s_2} v_{N-1}\int\cdots\int v_1(dt)^{N-1} & \cdots & \int_{s_{N-1}}^{s_N} v_{N-2}\int\cdots\int v_1\,(dt)^{N_1} \end{vmatrix}$$

$$= 0.$$

Let s_1 be viewed as a variable and denote it by s. Then, we see that $\Delta = \Delta(s)$ vanishes at $s = s_2$ and at $s = s_1$ by assumption. The mean value theorem says that there exists an s_1' in the interval (s_1, s_2) such that the determinant obtained from Δ by changing the first column by

$$\left(v_{N-1}(s_1'),\, v_{N-1}(s_1')\int_0^{s_1'} v_{N-2}\,dt,\, \ldots,\, v_{N-1}(s_1')\int_0^{s_1'} v_{N-2}\int\cdots\int v_1\,(dt)^{N-2}\right)^*$$

is equal to zero (* denotes the transposed vector). By the same arguments, we can find a sequence $s_1', s_2', \ldots, s_{N-1}'$ such that

$$\begin{vmatrix} 1 & 1 & \cdots & 1 \\ \int_0^{s_1'} v_{N-2}\,dt & \int^{s_2'} v_{N-2}\,dt & \cdots & \int_0^{s_{N-1}'} v_{N-2}\,dt \\ \cdots & \cdots & \cdots & \cdots \\ \int_0^{s_1'} v_{N-2}\int\cdots\int v_1\,(dt)^{N-2} & \int_0^{s_2'} v_{N-2}\int\cdots\int v_1\,(dt)^{N-2} & \cdots & \int_0^{s_{N-1}'} v_{N-2}\int\cdots\int v_1\,(dt)^{N-2} \end{vmatrix}$$

$$= 0.$$

where $s_2 < s_2' < s_3 < s_3' < \cdots < s_{N-1}' < s_N$. The last determinant is of the same type as the original Δ, so that the same argument can be applied to get a determinant of the same type and of lower order. Repeating this procedure, we finally come to a simple determinant.

$$\begin{vmatrix} 1 & 1 \\ \int_0^{s_1^{(N-2)}} v_1\,dt & \int_0^{s_2^{(N-2)}} v_1\,dt \end{vmatrix},$$

which is equal to zero; in other words $\int_{s_1^{(N-2)}}^{s_2^{(N-2)}} v_1\,dt = 0$, with $s_1^{(N-2)} < s_2^{(N-2)}$. While $v_1(t)$ is a continuous function on the interval $(s_1^{(N-2)}, s_2^{(N-2)})$ and never vanishes. This leads us to a contradiction.

(ii) Suppose that $\{g_i(u)\}$ is linearly dependent in $L^2([0, t])$ for some $t > 0$. Then there exist constants $a_1, a_2, \ldots, a_N$ that satisfy

$$\sum_{i=1}^{N} a_i g_i(u) = 0, \qquad u \in [0, t].$$

Dividing both sides by $v_i(u) \neq 0$ and taking the derivative, $i = 1, 2, \ldots, N$, successively, we finally arrive at the equation $v_N(t) \equiv 0$. This contradicts our assumption. □

We are now in a position to study the finer properties of multiple Markov Gaussian processes in the restricted sense by using the differential operator L_t. For the remainder of this section, a Gaussian process $X = \{X(t, \omega);\ t \in I\}$ is assumed to satisfy the assumptions (A.1), (A.2$'$), and (A.3$'$) the last assumption ensures separability of the Hilbert space $H(X)$.

DEFINITION 5.3. A Gaussian process $X = \{X(t, \omega);\ t \in I\}$ that satisfies (A.1), (A.2), and (A.3) is called an *N-ple Markov process in the restricted sense* if there exist a differential operator L_t of the form (5.20) for a system of functions v_i, $i = 1, 2, \ldots, N$, which satisfy (5.18) and (5.19); if $X(t)$ belongs to the domain of $L_t^{(1)}$, and

$$U(t) = L_t^{(1)} X(t) \tag{5.31}$$

is an additive process (that is, $U(t)$ has independent increments) with absolutely continuous variance.

The last requirement ensures the following expression:

$$\sigma^2(t)\ (= E(U(t)^2)) = \int_0^t v_0(u)^2\, du. \tag{5.32}$$

If a given Gaussian process X is an N-ple Markov process in the restricted sense, then the process $U = \{U(t);\ t \in I\}$ obtained by the formula (5.31) is expressible as a Wiener integral of the form

$$U(t) \int_0^t v_0(u)\, dB_0(u), \tag{5.33}$$

or equivalently as

$$dU(t) = v_0(t)\, dB_0(t) \quad \text{with } U(0) = 0. \tag{5.33'}$$

From this expression, we obtain a formal (but can be justified as before) differential equation

$$L_t X(t) = \dot{B}_0(t) \qquad \left(\dot{B}_0(t) = \frac{d}{dt} B_0(t)\right). \tag{5.34}$$

A solution to equation (5.34) with initial conditions $X(0) = X'(0) = \cdots = X^{(N-1)} = 0$ is

$$X(t) = \int_0^t R(t, u) \dot{B}_0(u)\, du, \tag{5.35}$$

where $R(t, u)$ is the Green's function given by (5.25). Such formal considerations suggest the observation below. In fact, the formula (5.35) becomes significant from the viewpoint of the canonical representation theory. To show this, we begin with a lemma.

LEMMA 5.4. *Let* $X = \{X(t);\ t \in I\}$ *be given by*

$$X(t) = \sum_{i=1}^{N} f_i(t) \int_0^t g_i(u)\, dB_0(u),$$

with

(i) $f_i(t) \in \mathbf{C}^1(I^0)$, $g_i(u) \in \mathbf{C}(I)$, $i = 1, 2, \ldots, N$,

and

(ii) $\sum_{i=1}^{N} f_i(t) g_i(t) \equiv 0$.

Then the derivative $X'(t)$ *exists in the mean square sense written in the form*

$$X'(t) = \sum_{i=1}^{N} f'_i(t) \int_0^t g_i(u)\, dB_0(u). \tag{5.36}$$

PROOF. For $h > 0$, we have

$$\begin{aligned} &X(t+h) - X(t) \\ &\quad = \sum_{i=1}^{N} f_i(t+h) \int_t^{t+h} g_i(u)\, dB_0(u) + \sum_{i=1}^{N} \{f_i(t+h) - f_i(t)\} \int_0^t g_i(u)\, dB_0(u). \end{aligned}$$

The order of the first term is

$$h \sum_{i=1}^{N} f'_i(t) g_i(t)(B_0(t+h) - B(t)) + o(h^2) = o(h).$$

Divide the second term by h and let $h \to 0$ to see the existence of the derivative $X'(t)$ as well as the explicit expression (5.36) for $X'(t)$. □

For the remainder of this section, we assume the differential operator L_t is formed by v_i, $i = 1, 2, \ldots, N$, which satisfy conditions (5.18) and (5.19).

THEOREM 5.2. (i) *An* N*-ple Markov Gaussian process* X *is uniquely determined by the operator* L_t *and a Brownian motion* $B_0 = \{B_0(t);\ t \in I\}$.

(ii) *The process* X *determined in* (i) *has canonical representation of the form*

$$X(t) = \int_0^t R(t, u)\, dB_0(u),$$

where the kernel $R(t, u)$ *is the Green's function for* L_t.

PROOF. (i) Define $U(t)$ by (5.33) by using the given v_0 and B_0. Then we can form the Green's function $R(t, u)$ for L_t using formula (5.25). Set

$$X(t) = \int_0^t R(t, u)\, dB_0(u), \tag{5.37}$$

then the process X satisfies equation (5.31), which is proved by Lemma 5.4. Thus, an N-ple Markov Gaussian process X has been formed from L_t and B_0.

For the proof of uniqueness, it suffices to show that

$$X(t) = 0 \quad \text{if } \frac{d}{dt}\frac{1}{v(t)}X(t) = 0, \qquad (v(t) \neq 0) \tag{5.38}$$

for a Gaussian process $X = \{X(t);\ t \in I\}$ satisfying conditions (A.1), (A.2$'$), and (A.3). In fact, if there were two processes, say X_1 and X_2 satisfying (5.31), then $L_t^{(1)}X(t) = 0$ must hold for $X(t) = X_1(t) - X_2(t)$. If (5.38) is true, then $L_2^{(2)}X(t) = 0$ is implied, where the second formula of (5.20) is used for $j = 2$. Repeating this argument N times, we prove that $X(t) = 0$ for $t \in I$. Hence, uniqueness follows.

Now we come to the proof of (5.38). The equation

$$\frac{d}{dt}\frac{1}{v(t)}X(t) = 0 \quad \text{implies} \quad X(t) = v(t)C,$$

where C is some constant random variable independent of t. However, $C = 0$ follows from (A.2$'$). Hence, $X(t) = 0$ has been derived.

The proof of (ii), that is, the canonical property of (5.37), is almost trivial, since Lemma 5.2 guarantees that the criterion stated in Theorem 4.4. is satisfied. □

The assertions below follows from the theorem above.

COROLLARY. *Suppose that* $X = \{X(t);\ t \in I\}$ *is an* N*-ple Markov Gaussian process in the restricted sense satisfing the properties expressed by* (5.31) *and* (5.32).

(i) *If* $v(t)$ *is a continuous function that never vanishes, then* $\{v(t)X(t, \omega);\ t \in I\}$ *is again an* N*-ple Markov Gaussian process in the restricted sense.*

(ii) $\{(d/dt)(1/v_0(t))X(t, \omega);\ t \in I\}$ *is an* $(N-1)$*-ple Markov Gaussian process in the restricted sense.*

REMARK. Here is an important remark regarding Theorem 5.2. The canonical kernel of an N-ple Markov Gaussian process in the restricted sense is the Green's function $R(t, u)$, which is formed in the well-known manner in analysis for the given operator L_t. Now one may ask how to construct the innovation. Indeed, it can be constructed in an exact form. Start with $L_t^{(1)}X(t) = U(t)$ and obtain $(dU(t)/v_0(t)dt) = \dot{B}_0(t)$, which is nothing but the innovation of the process X. Intuitively speaking, a canonical representation can be obtained by a method analogous to that of variation of constants in the theory of ordinary differential equations. This fact is a feature particular to a multiple Markov Gaussian process in the restricted sense.

The relation to the general multiple Markov property is stated in the following theorem.

THEOREM 5.3. *An N-ple Markov Gaussian process in the restricted sense is an N-ple Markov Gaussian process.*

PROOF. Assume that an N-ple Markov Gaussian process X is given by (5.37). Then the functions $f_i(t)$ and $g_i(t)$, $i = 1, 2, \ldots, N$, involved in the canonical kernel must satisfy properties (i) and (ii) in Lemma 5.3. The N-ple Markov property of X is therefore proved by applying Theorem 5.1. □

One can now raise the question as to when an N-ple Markov Gaussian process is so in the restricted sense. The answer to this question is given by the condition on the canonical kernel. Assume that the canonical kernel

$$F(t, u) = \sum_{i=1}^{N} f_i(t) g_i(u), \qquad u \leq t,$$

for an N-ple Markov Gaussian process satisfies the conditions

$$\begin{aligned} &f_i, g_i \in \mathbf{C}^\infty(I^0), \qquad i = 1, 2, \ldots, N, \\ &f_i(t) \neq 0, \qquad t \in I^0, \qquad i = 1, 2, \ldots, N, \\ &W(g_1, g_2, \ldots, g_i)(u) \neq 0, \qquad u \in I^0, \qquad i = 1, 2, \ldots, N, \end{aligned} \tag{5.39}$$

where $W(g_1, g_2, \ldots, g_i)(u)$ is the Wronskian of $g_1, g_2, \ldots, g_i$ evaluated at u. With this assumption we can prove the existence of $\mathbf{C}(I^0)$ functions $v_0, v_1, \ldots, v_{N-1}$ such that

$$\begin{aligned} g_i(u) = (-1)^{N-i} v_0(u) \int^u v_1(u_1)\, du_1 \int^{u_1} v_2(u_2)\, du_2 \int^{u_2} \\ \cdots \int^{u_{N-i-1}} v_{N-i}(u_{N-i})\, du_{N-i}, \\ i = 1, 2, \ldots, N. \end{aligned} \tag{5.40}$$

Moreover, the lower limit of the integrals can be taken to be 0 if the system $\{f_i\}$ is appropriately chosen using replacement by linear combinations. In addition, under the assumption (5.39), we see that the partial derivatives $F^{(i)}(t, u) = \partial^i F(t, u)/\partial t^i$, $t > u$, $i = 1, 2, \ldots, N-1$, exist.

THEOREM 5.4. *Suppose that an N-ple Markov Gaussian process $X = \{X(t, \omega);\ t \in I\}$ has an innovation B_0 and the canonical kernel $F(t, u) = \sum_{i=1}^N f_i(t) g_i(u)$, $u \leq t$, is formed by g_i satisfying assumption* (5.40).

(i) *If $F(t, t) = F^{(1)}(t, t) = \cdots = F^{(N-2)}(t, t) = 0$ and $F^{(N-1)}(t, t) \neq 0$, $t \in I^0$, then X is an N-ple Markov Gaussian process in the restricted sense.*

(ii) *If there is an integer $k \in \{1, 2, \ldots, N-2\}$ such that*

$$F(t, t) = F^{(1)}(t, t) = \cdots = F^{(k-1)}(t, t) = 0$$

and

$$F^{(k)}(t, t) \neq 0, \qquad t \in I^0,$$

then there exist an N-ple Markov Gaussian process $Y = \{Y(t, \omega);\ t \in I\}$ *in the restricted sense and a differential operator* M_t *of order* $N - k - 1$ *such that*

$$X(t, \omega) = M_t Y(t, \omega), \qquad t \in I^0. \tag{5.41}$$

Proof. (i) From the assumption, we claim that $X(t)$ is differentiable $N-1$ times. This follows from Lemma 5.4. Since $g_i(u)$, $i = 1, 2, \ldots, N$, are expressed in the form (5.40), the v_i $i = 1, 2, \ldots, N-1$, and a suitably chosen $v_N \in \mathbf{C}^\infty(I^0)$ determine a differential operator L_t. Noting that each v_i is in $\mathbf{C}^\infty(I^0)$, we show that L_t is a differential operator in the ordinary sense. Thus, our assumption implies that $F(t, u)$ is the Green's function for L_t. Using part (i) of the corollary to Theorem 5.2, we see that X is an N-ple Markov Gaussian process in the restricted sense.

(ii) As in the proof above, the differential operator L_t is also determined by the g_i, $i = 1, 2, \ldots, N$, and v_N. We denote the Green's function for the operator L_t by $R(t, u) = \sum_{i=1}^N f_i(t) g_i(u)$, $u \le t$. (The $\tilde{f}_i(t)$ is to be $f_i(t)$ in the notation of (5.22).) Take a differential operator

$$M_t = \sum_{k=0}^N a_k(t) \left(\frac{d}{dt}\right)^{N-k}$$

such that $M_t \tilde{f}_i = f_i$, $i = 1, 2, \ldots, N$. As is easily seen from the proof of Lemma 5.3, we have

$$\det \left(\frac{d^i}{dt^i} \tilde{f}_i(t_j) \right) \neq 0 \quad \text{a.e.}$$

Therefore, the ratios among $a_0(t), a_1(t), \ldots, a_N(t)$ are uniquely determined by the linear system

$$\sum_{k=0}^N a_k(t) \frac{d^{N-k}}{dt^{N-k}} \tilde{f}_i(t) = f_i(t), \qquad i = 1, 2, \ldots, N. \tag{5.42}$$

Multiply both sides by $g_i(t)$ and sum up all of them. Then we see that $a_0(t) \equiv 0$ from $F(t, t) = 0$ and from the properties of the Green's function: $R^{(k)}(t, t) = 0$, $k = 1, 2, \ldots, N-2$. Further, taking derivatives of both sides of (5.42), multiplying by $g_i(t)$, and then summing, we can see that $a_1(t) \equiv 0$ from $F^{(1)}(t, t) = 0$ and the properties of the Green's function. Repeating this argument $(k-1)$-times, we prove that M_t must be a differential operator of order $N - k - 1$. By the assumption on $F(t, u)$ and the requirement for $M_t : M_t \tilde{f}_i(t) = f_i(t)$, $i = 1, 2, \ldots, N$, we derive the desired equation (5.41). □

Example. A Gaussian process X with representation

$$X(t) = \int_0^t (3t^2 - 4tu + u^2)\, dB_0(u)$$

is a triple Markov Gaussian process, but not in the restricted sense, because X is differentiable only once. This is the case where $N = 3$ and $k = 1$ in Theorem 5.4 (ii). In fact, M_t can be taken to be $\frac{d}{dt}+1$ and the corresponding Y, which is a triple Markov Gaussian process in the restricted sense, is given by

$$Y(t) = \int_0^t (t-u)^2 \, dB_0(u).$$

As an application of the results obtained so far, we shall be concerned again with the prediction problem stated in §4.5. Remember that the best predictor is obtained from Proposition 4.2. Suppose that $X = \{X(t, \omega);\ t \in I\}$ is an N-ple Markov Gaussian process in the restricted sense and that the canonical kernel is expressed in the form $R(t, u) = \sum_{i=1}^N f_i(t) g_i(u)$ by using the functions $f_i(t)$, $g_i(u)$ $i = 1, 2, \ldots, N$, given by (5.22) and (5.23), respectively. The best predictor $\widehat{X}(t, s)$ has the form

$$\widehat{X}(t, s) = \int_0^s R(t, u)\, dB_0(u) = \sum_{i=1}^N f_i(t) \int_0^s g_i(u)\, dB_0(u). \tag{5.43}$$

For the purpose of applications, it is convenient to write the right-hand side of (5.43) in terms of the $\{X(u);\ u \le s\}$. To this end, we note the following fact. With the notation established in (5.20),

$$L_t^{(j)} f_j(t) = 0, \qquad i = 1, 2, \ldots, N-j.$$

Now, we introduce some additional notation

$$U_i(t) = \int_0^t g_i(u)\, dB_0(u)$$

and

$$f^{[k]}(t) = L_t^{(N-k)} f(t), \quad k = 1, 2, \ldots, N-1, \quad f^{[0]}(t) \equiv 0.$$

Then we can write

$$L_t^{(N-k)} X(t) = \sum_{i=k+1}^N f_i^{[k]}(t) U_i(t), \qquad k = 0, 1, \ldots, N-1. \tag{5.44}$$

Since $\prod_{k=1}^N f_k^{[k-1]}(t) = 1 \neq 0$, $t \in I^0$, equation (5.44) is solved in terms of the $U_i(t)$, and the solution is of the form

$$U_i(t) = \sum_{k=0}^{N-1} a_{ik}(t) L_t^{(N-k)} X(t). \tag{5.45}$$

From (5.43) and (5.45), we have

$$\widehat{X}(t, s) = \sum_{k=0}^{N-1} b_k(t, s) L_s^{(N-k)} X(s) \quad \text{with } b_k(t, s) = \sum_{i=1}^N f_i(t) a_{ik}(s). \tag{5.46}$$

Thus, the best predictor is explicitly expressed as a linear combination of $X(s)$, $X'(s)$, ..., $X^{(N-1)}(s)$. The essential point is that the best predictor is constructed "locally" in time, depending upon the values in the neighbourhood of s (actually only the values at $u \le s$), i.e., "causally".

As the next problem, consider the best predictor for a general N-ple Markov Gaussian process X. Let X be a process given in Theorem 5.4 (ii). Then $Y(t)$ defined by (5.41) is expressed as a linear functional of $\{X(u);\ u \le t\}$ by the inverse operator M_t^{-1}, which is an integral operator. In addition,

$$\mathbf{B}_t(X) = \mathbf{B}_t(Y) \tag{5.47}$$

holds for every $t \in I$. (This property will be clarified in the next section.) Applying the operator in (5.45) to $Y(t)$, the best predictor $\widehat{X}(t, s)$ is eventually evaluated in the following manner:

$$\begin{aligned}\widehat{X}(t, s) &= \sum_{j=1}^{N} f_i(t) U_i(s) = \sum_{j=1}^{N} f_i(t) \left(\sum_{k=0}^{N-1} a_{ik}(s) L_s^{(N-k)} Y(s) \right) \\ &= \sum_{k=0}^{N-1} b_k(t, s) L_s^{(N-k)} (M_s^{-1} X(s)).\end{aligned} \tag{5.48}$$

In this case, the best predictor can not be constructed "locally" in time, since M_s^{-1} appears, which is not a local operator but a "causal" operator. Such a predictor involves a nonlocal statistics extended over the entire history.

§4. Multiple Markov stationary Gaussian processes

The class of stationary multiple Markov Gaussian processes occupies a particularly important place in the collection of multiple Markov Gaussian processes. We shall, in this section, discuss the canonical representation of such stationary processes by using the general theory developed in Chapter III. The first advantage of specifying the processes to be a stationary multiple Markov process is that methods to obtain the canonical kernel and the innovation are available in an exact manner. The second is that canonical and noncanonical representations can be discriminated in an explicit way. Indeed, the canonical kernel of the multiple Markov stationary Gaussian process is formed explicitly by the particular factorization of the spectral density function. The third advantage is that one is able to see the possibility of constructing infinitely multiple Markov Gaussian process as a generalization of multiple Markov processes in the restricted sense.

Returning to the second advantage, we note that P. Lévy's original idea for a theory of canonical representations, which is interesting in the historical viewpoint, arose in this context. As is well known he began the theory with an investigation of the canonical representation of the $M(t)$-process which is the average of a Lévy's Brownian motion over the sphere with radius t. If the parameter space of the Lévy's Brownian motion is odd dimensional, then the $M(t)$-process is a typical example of a multiple Markov Gaussian process which is transformed into a stationary process by a simple change of time scale.

In this section, the parameter space is taken to be $I = (-\infty, \infty)$. The process $X = \{X(t);\ t \in I\}$ in question is assumed to be a complex stationary Gaussian process satisfying assumptions (A.1), (A.2$'$), and (A.3$'$), which imply that the covariance function of X, $\gamma(h) = E[X(t+h)\overline{X(t)}]$, is independent of t and continuous in h. Using the results in Chapter III, the spectral representation of $\gamma(h)$ and that of $X(t)$ are expressed, respectively, in the following manner:

$$\gamma(h) = \int e^{ih\lambda} f(\lambda)\, d\lambda, \tag{5.49}$$

and

$$X(t) = \int e^{it\lambda}\, dZ(\lambda), \qquad E(|dZ(\lambda)|^2) = f(\lambda)\, d\lambda. \tag{5.50}$$

There exists a function $G(u)$ such that $G(u) = 0$ for $u \in (-\infty, 0]$, the Fourier inverse transform $\widehat{G}(\lambda)$ of $G(u)$ satisfies $|\widehat{G}(\lambda)|^2/2\pi = f(\lambda)$, and the X itself is canonically represented in the form

$$X(t) = \int_{-\infty}^{t} G(t-u)\, dB(u), \tag{5.51}$$

by virtue of Theorem 3.4 in Chapter III.

Keeping this situation, we further assume the Markov property. To fix ideas, we first observe a simple Markov Gaussian process. By Theorem 5.1, the canonical kernel $G(t-u)$ is written in the form

$$G(t-u) = f(t)g(u), \qquad u \leq t.$$

From the measurability of g, it follows easily that f is measurable. We therefore conclude that f, g, and also G are all exponential functions. Set $g(u) = ce^{\lambda u}$, with constants c and λ. Since g must be in $L^2((-\infty, t])$ for any t, λ must be positive. We now have $f(t) = c'e^{-\lambda t}$ with a constant c', and thus $G(t-u) = cc'e^{-\lambda(t-u)}$, $\lambda > 0$ is obtained. By setting $a = cc'$, (5.51) is expressed in the form

$$X(t) = ae^{-\lambda t} \int_{-\infty}^{t} e^{\lambda u}\, dB(u). \tag{5.52}$$

The process $X = \{X(t);\ t \in I\}$ given by (5.52) clearly satisfies a stochastic differential equation

$$dX(t) = -\lambda X(t)\, dt + a\, dB(t), \tag{5.53}$$

which is called the *Langevin equation*. Equation (5.53) tells us that X is the well-known *Ornstein-Uhlenbeck Brownian motion*.

The observation above shows that a stationary Gaussian process with the simple Markov property is a particular and, in fact, very important process.

Now the following question naturally arises: what is a stationary Gaussian process with multiple Markov property? To answer this question we first recall that the multiple Markov property should be defined as a natural generalization of the simple Markov property; hence, the canonical kernel is also expected to be a generalization of the exponential function as in (5.52). These considerations lead to the next Lemma.

LEMMA 5.5. *If a Goursat kernel*

$$F(t, u) = \sum_{i=1}^{N} f_i(t) g_i(u), \qquad u \le t, \tag{5.54}$$

(of order N) is a function only of $t-u$, then $\{f_i(t);\ i = 1, 2, \dots, N\}$ is a fundamental system of solutions of a linear ordinary differential equation with constant coefficients, while $\{g_i(u);\ i = 1, 2, \dots, N\}$ is a fundamental system of solutions of the adjoint differential equation.

REMARK. This lemma shows that the canonical kernel $F(t, u) = F(t-u)$ for a stationary N-ple Markov Gaussian process is a Goursat kernel of order N expressed as a linear combination of functions of the form

$$(t-u)^k e^{-\mu(t-u)}, \qquad u \le t, \text{ to } k = 0, 1, \dots, n \ (\le N), \tag{5.55}$$

μ being a constant with $\operatorname{Re} \mu > 0$.

PROOF OF LEMMA 5.5. Consider $F(t-u)$ in the domain $D_0 = \{(u, t);\ u \le 0,\ t \ge 0\}$, and let $\mathscr{D}_0$ be the subspace of the Schwartz space $\mathscr{D}$, involving all $\mathbf{C}^\infty$-functions whose supports are in $(-\infty, 0]$. If $\varphi \in \mathscr{D}_0$, then the convolution

$$(F * \varphi)(t) = \int_{-\infty}^{\infty} F(t-u)\varphi(u)\, du,$$

is a $\mathbf{C}^\infty$-function over $[0, \infty)$. Next, we shall show that there exist functions $\varphi_1, \varphi_2, \dots, \varphi_N$ such that

$$\det((g_i, \varphi_j))_{i, j=1,2,\dots,N} \ne 0, \tag{5.56}$$

where, the bracket $(\ ,\ )$ means the inner product in $L^2(I)$. First, we note that there exists $\varphi_1 \in \mathscr{D}_0$ such that $(g_1, \varphi_1) \ne 0$. In fact, if there were no such function, then g_1 would be zero almost everywhere. We now use mathematical induction to prove the existence of the system $\{\varphi_j\}$ satisfying (5.56). Assume that $\varphi_1, \varphi_2, \dots, \varphi_n$ $(n < N)$ are obtained and satisfy

$$\det((g_i, \varphi_j))_{i, j=1,2,\dots,n} \ne 0.$$

Suppose that the determinant

$$\begin{vmatrix} (g_1, \varphi_1) & (g_1, \varphi_2) & \cdots & (g_1, \varphi_n) & (g_1, \varphi) \\ (g_2, \varphi_1) & (g_2, \varphi_2) & \cdots & (g_2, \varphi_n) & (g_2, \varphi) \\ \cdots & \cdots & \cdots & \cdots & \cdots \\ (g_{n+1}, \varphi_1) & (g_{n+1}, \varphi_2) & \cdots & (g_{n+1}, \varphi_n) & (g_{n+1}, \varphi) \end{vmatrix}$$

is equal to zero for all $\varphi \in \mathscr{D}_0$. Then, the determinant obtained by replacing the last column (g_j, φ), $j = 1, 2, \dots, n+1$, by $g_j(u)$, $j = 1, 2, \dots, n+1$, is still zero for almost all $u \le 0$. Noting that the value of the determinant is a linear combination of $\{g_i(u); i = 1, 2, \dots, n+1\}$, the last conclusion is inconsistent with the property that the g_i, $i = 1, 2, \dots, N$, are linearly independent. Therefore, there exists a function φ for which the determinant is not zero. This φ is the sought for φ_{n+1}. Hence, we have proved (5.56).

We then form a function $(F * \varphi_j)(t)$ using $\{f_i\}$ and $\{g_i\}$,

$$(F * \varphi_j)(t) = \sum_{i=1}^{N} (g_i, \varphi_j) f_i(t), \qquad j = 1, 2, \dots, N,$$

the left-hand side of which is a $\mathbf{C}^\infty$-function of $t \ge 0$. By (5.56), each $f_i(t)$ is expressible as a linear combination of the $(F * \varphi_j)(t)$, $1 \le j \le N$ so that the $f_i(t)$ are elements of $\mathbf{C}^\infty((0, \infty))$. Noting that $F(t) = \sum_{i=1}^N f_i(t) g_i(0)$, $t > 0$, we see that $F \in \mathbf{C}^\infty((0, \infty))$. Using this property and (5.11), we can prove that each $g_i(u)$ belongs to $\mathbf{C}^\infty(-\infty, 0))$.

Repeating the same argument for the domains $D_a = \{(u, t); u \le t, t \ge a\}$, $a \in I$, we can finally conclude that

$$f_i, g_i \in \mathbf{C}^\infty(I), \qquad i = 1, 2, \dots, N.$$

The last step in the proof is concerned with the differential equation satisfied by the f_i. Observing the equalities

$$\begin{aligned}\sum_{i=1}^{N} f_i^{(k)}(t) g_i(u) &= \frac{\partial^k}{\partial t^k} F(t-u) \\ &= (-1)^k \frac{\partial^k}{\partial u^k} F(t-u) \\ &= (-1)^k \sum_{i=1}^{N} f_i(t) g_i^{(k)}(u), \qquad k = 1, 2, \dots, N, \quad u < t,\end{aligned} \tag{5.57}$$

we know that the $F^{(k)}(t)$, $k = 0, 1, 2, \dots, N$, are linearly dependent. In fact, the $F^{(k)}(t)$ are all linear combinations of $f_i(t)$, $i = 1, 2, \dots, N$, as is seen by setting $u = 0$ in the formula (5.57). Hence, there exist constants $b_0, b_1, b_2, \dots, b_N$ such that

$$L_t F(t) \equiv \sum_{j=0}^{N} b_j F^{(N-i)}(t) = 0, \qquad t > 0.$$

Replacing t with $t - u$ $(u < t)$, we have F expressed by the $f_i^{(k)}$ with coefficients the g_i. Since the g_i, $i = 1, 2, \dots, N$, are linearly independent, the differential equations

$$L_t f_i(t) = 0, \qquad t \in I, \; i = 1, 2, \dots, N,$$

must hold. Since $\{f_i\}$ is a linearly independent system satisfying $L_t f_i(t) = 0$, $i = 1, 2, \ldots, N$, for the ordinary differential operator L_t of order N, $\{f_i\}$ is the fundamental system of solutions to the differential equation $L_t f = 0$.

The result for the system $\{g_i\}$ can be proved in an analogous manner by using (5.57). □

THEOREM 5.5. *Let $X = \{X(t, \omega);\ t \in I\}$ be a complex stationary Gaussian process satisfying the assumptions* (A.1), (A.2′), *and* (A.3′). *If X is an N-ple Markov process, the canonical kernel is a Goursat kernel of order N, which is expressed as a linear combination of the functions listed in* (5.55).

PROOF. We know by Theorem 3.4 that X has a canonical representation and that the kernel is a function only of $t-u : F(t-u)$. In addition, by Theorem 5.1 $F(t-u)$ is a Goursat kernel of order N. Hence, $F(t-u)$ must be a linear combination of the functions listed in (5.55) of Lemma 5.5. □

REMARK. Theorem 5.5 is really a powerful result. With slight modification, the theorem is applied to a weakly stationary process. The conditional expectation $E(X(t_i)|\mathbf{B}_{t_0}(X))$ in Definition 5.2 is replaced by $U(t_0, t_i) \equiv$ the projection of $X(t_i)$ to $H_{t_0}(X)$. If the $U(t_0, t_i)$, $i = 1, 2, \ldots, N$, satisfy the conditions required there, then the kernel $F(t-u)$ of the moving average representation is also a function as in Theorem 5.5.

COROLLARY. *Let X be a Gaussian process given in Theorem 5.5. Then the inverse Fourier transform $c(\lambda)$ of its canonical kernel F*

$$c(\lambda) = \frac{1}{\sqrt{2\pi}} \int_0^\infty e^{-i\lambda t} F(t)\, dt$$

is a rational function of $i\lambda$, being expressible as

$$c(\lambda) = Q(i\lambda)/P(i\lambda), \qquad -\infty < \lambda < \infty, \tag{5.58}$$

where $P(z)$ and $Q(z)$ are polynomials in $z \in \mathbf{C}$ having no roots in the lower half plane such that the degree of $P = N$, and the degree of $Q < N$.

PROOF. Take $F(t-u)$ to be one of the functions given by (5.55). Then the Fourier inverse transform of $F(t)$ is $(i\lambda + \mu)^{-k-1}$, $\operatorname{Re}\mu > 0$, up to a constant. The function $c(\lambda)$ is a linear combination of such functions. Noting that $F(t-u)$ is a canonical kernel, the conclusion of Corollary follows from the discussion in §3.4. □

EXAMPLE 1. Take a particular function in the class of functions listed in (5.55). For example, let $\lambda = 1$ and $n = 1$. Then we have a Gaussian process X given by

$$X(t) = \int_{-\infty}^{t} (t-u)e^{-(t-u)}\, dB_0(u),$$

which is a double Markov process. Here B_0 may be a real or complex Brownian motion. The kernel is obviously canonical and

$$c(\lambda) = \frac{1}{\sqrt{2\pi}} \frac{1}{(1+i\lambda)^2}, \qquad f(\lambda) = \frac{1}{2\pi} \frac{1}{(1+\lambda^2)^2}.$$

EXAMPLE 2. Take $n = 0$ and let μ be 1 and 2 in (5.55). With these choices, we define a stationary Gaussian process X by

$$X(t) = \int_{-\infty}^{t} \{2e^{-(t-u)} - e^{-2(t-u)}\} \, dB_0(u), \qquad t \in R.$$

The kernel in $\{\ \}$ is clearly canonical and

$$c(\lambda) = \frac{1}{\sqrt{2\pi}} \frac{3+i\lambda}{(1+i\lambda)(2+i\lambda)}, \qquad f(\lambda) = \frac{1}{2\pi} \frac{9+\lambda^2}{(1+\lambda^2)(4+\lambda^2)}.$$

The examples above are both double Markov Gaussian processes and the representations are canonical. There is, however, a crucial difference between them; the first one is differentiable and hence it is double Markov in the restricted sense, while the second one is not.

We wish to call attention to Theorem 5.4 (ii) in connection with Example 2, where $N = 2$, $F(t, t) \equiv 1$, and $k = 0$. There is a Gaussian process Y, which is double Markov in the restricted sense and the Fourier inverse transform $c_Y(\lambda)$ of its canonical kernel is given by

$$c_Y(\lambda) = \frac{1}{\sqrt{2\pi}} \frac{1}{(1+i\lambda)(2+i\lambda)}.$$

The relation (5.41) that connects $X(t)$ and $Y(t)$ is given explicitly by

$$X(t) = e^{-3t} \frac{d}{dt} e^{3t} Y(t).$$

The recipe for calculating the best predictor of the process X is as follows. First one obtains $Y(t)$ from $X(s)$, $s \le t$, by the integral

$$Y(t) = e^{-3t} \int_{-\infty}^{t} e^{3s} X(s) \, ds,$$

then formula (5.48) is applied.

Incidentally, it is easily seen that the process $Y = \{Y(t, \omega);\ t \in I\}$ formed above satisfies the following stochastic differential equation

$$\frac{d^2}{dt^2} Y(t) + 3 \frac{d}{dt} Y(t) + 2Y(t) = \dot{B}_0(t).$$

Letting the coefficients be generalized, we obtain

$$m \frac{d^2}{dt^2} Y(t) + \beta \frac{d}{dt} Y(t) + \alpha Y(t) = \dot{B}_0(t).$$

The solution to this equation describes the dynamics of a harmonic oscillator having friction and random impluse. The constants m, β, and α denote the mass, the friction coefficient, and the elastic force, respectively. The term $\dot{B}_0(t)$ stands for a random shock at instant t.

Motivated by the above discussion on double Markov Gaussian processes, one can investigate the structure of a general N-ple Markov stationary Gaussian process X. A rational function $c(\lambda)$ of the form (5.58), up to constant, is uniquely associated with an N-ple Markov stationary Gaussian process. And there exists a unique Gaussian process $Y = \{Y(t, \omega);\ t \in I\}$ with $c(\lambda) = 1/P(i\lambda)$, and the process Y is an N-ple Markov Gaussian process in the restricted sense. If we write $Q(i\lambda) = c\prod_{j=1}^{M}(\lambda_j + i\lambda)$, with $\lambda_1 > \lambda_2 > \cdots > \lambda_M$, then the relation of X and Y is given by

$$X(t) = c\prod_{j=1}^{M}\left\{e^{-\lambda_j t}\frac{d}{dt}e^{\lambda_j t}\right\}Y(t) \tag{5.59}$$

or alternatively

$$\begin{aligned} Y(t) = c^{-1}e^{-\lambda_1 t}\int_{-\infty}^{t} e^{(\lambda_1-\lambda_2)s_1}\,ds_1\int_{-\infty}^{s_1} e^{(\lambda_2-\lambda_3)s_2}\,ds_2 \\ \cdots\int_{-\infty}^{s_{M-2}} e^{(\lambda_{M-1}-\lambda_M)s_{M-1}}X(s_{M-1})\,ds_{M-1}. \end{aligned} \tag{5.60}$$

It is often convenient to write the above relations in the form

$$X(t) = Q\left(\frac{d}{dt}\right)Y(t) \quad \text{or} \quad Y(t) = Q\left(\frac{d}{dt}\right)^{-1}X(t), \tag{5.61}$$

respectively, where one should note that the operator $Q(d/dt)^{-1}$ is causal, that is, $Y(t) \in H_t(X)$ for all t.

The differential operator L_t associated with the N-ple Markov Gaussian process Y is given by

$$L_t = P\left(\frac{d}{dt}\right), \tag{5.62}$$

where P is the polynomial appearing in formula (5.58). By using this operator and $Q(d/dt)$, equation (5.61) can be rephrased as a stochastic integro-differential equation

$$P\left(\frac{d}{dt}\right)Q\left(\frac{d}{dt}\right)^{-1}X(t) = \dot{B}_0(t). \tag{5.63}$$

Summing up, we can claim that an N-ple Markov stationary Gaussian process X is characterized as the solution to the equation (5.63), where P and Q are polynomials determined by (5.58). As for the canonical representation, we can say that $B_0 = \{B_0(t, \omega);\ t \in I\}$ is the innovation and that the canonical kernel $F(t-u)$ is obtained by

$$F(t-u) = Q\left(\frac{d}{dt}\right)R(t-u), \tag{5.64}$$

where $R(t-u)$ is Green's function for the differential operator L_t. With these tools some additional fine points about X can be discussed in an analytic manner, although we do not go into details here.

In particular, we note that if $c(\lambda) = k/P(i\lambda)$ with k a constant in the expression (5.63), then (5.63) is turned into

$$k^{-1}P\left(\frac{d}{dt}\right)X(t) = \dot{B}_0(t),$$

and the given process is an N-ple Markov Gaussian process in the restricted sense. Such a process is in agreement with the one discussed by J. L. Doob (1944).

The methods we have used so far are applicable also to some classes of nonstationary Gaussian processes. An important example of such a class involves a Gaussian process which can be transformed to a stationary process by a change of time scale. The rest of this section is devoted to this topic.

Let $\widetilde{X} = \{\widetilde{X}(t, \omega);\ t \in I\}$ be an N-ple Markov stationary Gaussian process. If we define a process X by

$$X(t) = \sqrt{t}\widetilde{X}\left(\frac{\log t}{2}\right), \tag{5.65}$$

then it is easy to see that $X = \{X(t, \omega);\ t \geq 0\}$ is again an N-ple Markov Gaussian process and that the transformation preserves differentiability.

We recall that $\widetilde{X}$ is a linear combination of Gaussian processes of the type $\int_{-\infty}^{t}(t-u)^k e^{-\lambda(t-u)}\,d\widetilde{B}_0(u)$, so that the transformation (5.65) gives us new Gaussian processes of the type

$$2^{-(k+1)/2}\int_0^t \left(\log\frac{t}{u}\right)^k \left(\frac{u}{t}\right)^{(\lambda-1)/2} dB_0(u). \tag{5.66}$$

Here both $\widetilde{B}_0$ and B_0 are (real or complex) Brownian motions.

If we study the integrand of (5.66), the following proposition seems to be natural. The proof is straightforward.

PROPOSITION 5.2. *If $X = \{X(t);\ t \geq 0\}$ has canonical representation with canonical kernel $F(t, u)$ which is a homogeneous function of order α, then the process $\widetilde{X} = \{\widetilde{X}(t);\ t \in I\}$ defined by*

$$\widetilde{X}(t) = e^{-(2\alpha+1)t}X(e^{2t})$$

is a stationary Gaussian process. Moreover, if X has the multiple Markov property, then so does $\widetilde{X}$.

§5. Lévy's $M(t)$-process

So far the structure of multiple Markov Gaussian processes, from the viewpoint of representation theory in terms of Brownian motion, has been discussed and investigated. Once again, in this section, we deal with Lévy's

$M(t)$-process, which provided motivation for our study of the multiple Markov property. We will revisit the ideas of the multiple Markov property as well as its generalization.

Lévy's $M(t)$-processes were defined from multidimensional parameter Brownian motion (see Chapter IV, §4.4, part E, Definition 4.4, and also the definition reviewed below).

A Gaussian system $X = \{X(A);\ A \in R^N\}$ is a *Lévy Brownian motion with R^n parameter* if the following conditions (i), (ii), and (iii) are satisfied:

(i) $E(X_N(A)) = 0,\ A \in R^N$,

(ii) $X(O) = 0,\ O$ being the origin of R^N,

(iii) $E(X(A) - X(B))^2 = r(A, B)$.

The covariance function Γ is

$$\Gamma(A, B) = E(X(A)X(B)) = \frac{1}{2}\{r(O, A) + r(O, B) - r(A, B)\}.$$

It can be proved that $X(A)$ is a continuous function of A almost surely. As is easily seen, X is an ordinary Brownian motion B_0 if $N = 1$. There are many plausible reasons why the above X may be called a Brownian motion, i.e., a multidimensional parameter generalization of ordinary Brownian motion, however no attempt is made here to explain this topic.

A Brownian motion X with multidimensional parameter has a much more intricate and in fact interesting probabilistic structure compared to ordinary Brownian motion. In particular, we can see a significant difference from the case $N = 1$, if we observe the Markov property. The following discussion on the so-called Lévy's *$M(t)$-process* may serve to illustrate the property.

Let $S_N(t)$ be the sphere in R^N with center at the origin and with radius t, and let σ_t be the uniform probability measure on $S_N(t)$: $\sigma_t(S_N(t)) = 1$. Let a Gaussian process $M_N = \{M_N(t);\ t \geq 0\}$ be defined by

$$M_N(t) = \int_{S_N(t)} X(A)\sigma_t(dA), \tag{5.68}$$

which is called the (Lévy's) *$M_N(t)$-process*. If N is not specified, it is simply called an *$M(t)$-process*.

Obviously $E(M_N(t)) = 0$, and the covariance function Γ_N is

$$\Gamma_N(t, s) = E(M_N(t)M_N(s)) = \frac{1}{2}(t + s - \rho_N(t, s)), \tag{5.69}$$

where

$$\rho_N(t, s) = \int_{A\in S_N(t)} \int_{B\in S_N(s)} r(A, B)\sigma_t(dA)\sigma_s(dB). \tag{5.70}$$

It seems to be difficult to be given an explicit form of Γ_N, but in the case $t = s$, it is easy to calculate

$$\rho_N(t, t) = (J_{N-2}/I_{N-2})t, \qquad t \geq 0 \tag{5.71}$$

for $N \geq 3$, where I_k and J_k are constants given by

$$I_k = \int_0^{\pi/2} \sin^k \theta \, d\theta \quad \text{and} \quad J_k = \int_0^{\pi} \sin^k \theta \sin(\theta/2) \, d\theta .$$

For the general case $s \neq t$, (5.70) is expressed in the form

$$\rho_N(t, s) = \frac{1}{2I_{N-2}} \int_0^{\pi} r \sin^{N-2} \theta \, d\theta \quad \text{with } r = (t^2 + s^2 - 2ts\cos\theta)^{1/2} . \tag{5.72}$$

The formula (5.72) tells us that in the case where N is even ρ_N becomes an elliptic integral, while in the case where N is odd, say $N = 2p+1$, ρ_N can be obtained explicitly by an elementary method. For $t > s > 0$, we obtain, from (5.69), the following results:

$$\Gamma_3(t, s) = \frac{s}{2} - \frac{s^2}{6t},$$

$$\Gamma_5(t, s) = \frac{s}{2} - \frac{s^2}{5t} + \frac{s^4}{70t^3},$$

$$\Gamma_7(t, s) = \frac{s}{2} - \frac{3s^2}{14t} + \frac{s^4}{42t^s} - \frac{s^6}{462t^5},$$

$$\Gamma_9(t, s) = \frac{s}{2} - \frac{2s^2}{9t} + \frac{s^4}{33t^3} - \frac{2s^6}{429t^5} + \frac{s^8}{2574t^7},$$

and so forth. We are now able to guess a general rule for defining Γ_N for N odd. The expression of $\Gamma_{2p+1}(t, s)$ always starts with $s/2$, then follows $s \times$ (polynomial in s/t of degree $2p-1$). Moreover, as the dimension N increases by 2, the degree of polynomial increases by 1. It seems to be difficult to find a general rule for the coefficients.

From the formula for Γ_N in terms of a polynomial, P. Lévy (1956) expected that the canonical kernel of the $M_{2p+1}(t)$-process is a polynomial in u/t. He succeeded in obtaining the canonical representation of $M_{2p+1}(t)$. The result can be summarized as follows. Let P_{2p+1} be a polynomial of order $2p-1$ given by

$$P_{2p+1}(u) = \frac{2p}{\sqrt{\pi}} \sqrt{I_{2p}} \int_u^1 (1 - x^2)^{p-1} \, dx .$$

Then the canonical representation of $M_{2p+1}(t)$ can be written as

$$M_{2p+1}(t) = \int_0^t P_{2p+1}\left(\frac{u}{t}\right) dB_0(u) . \tag{5.73}$$

For instance, in the cases of $p = 2, 3$, we have

$$M_5(t) = \sqrt{3} \int_0^t \left(\frac{2}{3} - \frac{u}{t} + \frac{u^3}{3t^3} \right) dB_0(u),$$

$$M_7(t) = \sqrt{10} \int_0^t \left(\frac{2}{5} - \frac{3u}{4t} + \frac{u^3}{2t^3} - \frac{3u^5}{20t^5} \right) dB_0(u) .$$

It is easy to check that these are canonical representations and that the covariances calculated from these representations are both in agreement with the $\Gamma_5(t, s)$ and $\Gamma_7(t, s)$ obtained previously.

Unfortunately it is almost hopeless to find the canonical representation for N even by the above method. Nevertheless, we can speak of some properties of the canonical kernel such as homogeneous function of t and u of order 0. It is therefore possible to appeal to Proposition 5.2 in order to make the $M(t)$-process stationary, so that the theory established in Chapter III can be applied. In fact, this idea suggested to us an approach to this investigation in the case where N is even.

Take the $M_N(t)$-process in question and form a stationary Gaussian process $X_N = \{X_N(t);\ t \in R\}$ by a change of time scale of $M_N(t)$,

$$X_N(t) = e^{-t} M_N(e^{2t}), \qquad t \in R. \tag{5.74}$$

The covariance function $\gamma_N(h)$ of X_N is easily obtained from the formulas (5.69) and (5.70),

$$\gamma_N(h) = \frac{1}{2}\left\{2\cosh|h| - \frac{1}{\sqrt{2}I_{N-2}}\int_0^\pi (\cosh(2|h|) - \cos\theta)^{1/2} \sin^{N-2}\theta\, d\theta\right\}. \tag{5.75}$$

Lemma 5.6. *For $N \geq 4$, the covariance function $\gamma_N(h)$, $h \in R$, is twice continuously differentiable and satisfies the following equation:*

$$(2N-3)^2\gamma_N(h) - \gamma_N''(h) = 4(N-1)(N-2)\gamma_{N-2}(h). \tag{5.76}$$

Proof. Observing the expression (5.75), we see that γ is twice differentiable in h under the integral sign. With this property in mind the equation (5.76) is obtained by a simple calculation. □

Equation (5.76) gives a relationship between X_N and X_{N-2}, as is shown by

Theorem 5.5. *The stationary Gaussian processes X_N and X_{N-2} have canonical representations expressed by the common innovation process $B_0 = \{B_0(t, \omega);\ t \in R\}$. With such representations they satisfy the equation*

$$e^{-(2N-3)t}\frac{d}{dt}e^{(2N-3)t}X_N(t) = c_N X_{N-2}(t), \tag{5.77}$$

for $N \geq 4$, where $c_N = 2\sqrt{(N-1)(N-2)}$.

Proof. The stationary Gaussian processes in question satisfy conditions (A.1), (A.2′), and (A.3). Hence, we can form canonical representations using a common Brownian motion B_0 by using the result in Chapter IV.

Let the spectral representation of X_N be given by

$$X_N(t) = \int e^{it\lambda}\, dZ_N(\lambda).$$

Since X_N is purely nondeterministic, the spectral measure should be absolutely continuous: $E(|dZ_N(\lambda)|^2) = f_N(\lambda)\,d\lambda$. Differentiability of X_N implies $\int \lambda^2 f_N(\lambda)\,d\lambda < \infty$. Noting these properties, we see that the process given by the left-hand side of (5.77) is expressed in the form $\int (i\lambda + 2N-3)e^{it\lambda}\,dZ_N(\lambda)$ and has covariance function

$$\int e^{ih\lambda}(\lambda^2 + (2N-3)^2) f_N(\lambda)\,d\lambda = -\gamma_N''(h) + (2N-3)^2\gamma_N(h).$$

By Lemma 5.6, this is equal to $c_N^2\gamma_{N-2}(h)$. Since X_N and X_{N-2} are represented by the same innovation $\dot{B}_0$, the relation (5.77) is derived immediately. □

This theorem means that the investigation of the canonical representation and the Markov property of the $M_N(t)$-process may be reduced to those of X_1 or X_2. In what follows we shall discuss those properties in this vein.

(1°) In the case $N = 2p+1$, denote the differential operator $c_N^{-1}e^{-(2N-3)t}(d/dt)e^{(2N-3)t}$ by D_N and apply it to $X(t)$ to get (5.57). Repeating this operation, we finally come to the expression

$$D_3 D_5 \cdots D_{2p+1} X_{2p+1}(t) = X_1(t). \tag{5.78}$$

The processes M_1 and X_1 are, up to a constant, ordinary Brownian motion and an Ornstein-Uhlenbeck process, respectively. The equation above can be used to confirm that X_{2p+1} is a $(p+1)$-ple Markov Gaussian process in the restricted sense. Additional results regarding the multiple Markov property may be observed from the following computation of the spectral density function. Namely, for $X_{2p+1}(t)$ we have

$$f_{2p+1}(\lambda) = \frac{c_{2p+1}\cdot c_{2p-1}\cdots c_3\cdot c_1}{((4p-1)^2+\lambda^2)((4p-5)^2+\lambda^2)\cdots(3^2+\lambda^2)(1+\lambda^2)},$$

where c_N, $N \geq 3$, are constants given in (5.77) and $c_1 = 1/2$.

In order to proceed to the canonical kernel, we now recall the relation between the spectral density function and the kernel. The Fourier inverse transform $c(\lambda) = c_{2p+1}(\lambda)$ of the canonical kernel is obtained by the factorization of the spectral density function (See §3.4 and §5.4). It must have the form

$$c_{2p+1}(\lambda) = \frac{c_{2p+1}^{1/2}\cdot c_{2p-1}^{1/2}\cdots c_3^{1/2}}{(4p-1+i\lambda)(4p-5+i\lambda)\cdots(3+i\lambda)}\cdot\frac{c_1^{1/2}}{1+i\lambda}. \tag{5.79}$$

Indeed, $c_{2p+1}(\lambda)$ has to be a function such that $|c_{2p+1}(\lambda)|^2/2\pi = f_{2p+1}(\lambda)$ and such that $c_{2p+1}(w)$ is regular in the lower half plane. By expanding into partial fractions and then taking the Fourier transform term by term, we have the canonical kernel and hence the canonical representation of $X_{2p+1}(t)$ in the following form:

$$X_{2p+1}(t) = \int_{-\infty}^{t}\left\{a_0 e^{-(t-u)} + \sum_{k=1}^{p} a_k e^{-(4k-1)(t-u)}\right\} d\widetilde{B}_0(u), \tag{5.80}$$

where a_k, $k = 0, 1, \ldots, p$, are constants and where $\widetilde{B}_0$ is a Brownian motion. Moreover, by the time change which is inverse to (5.74), we come to the canonical representation of M_{2p+1}:

$$M_{2p+1}(t) = \int_0^t \left(\frac{1}{\sqrt{2}} a_0 + \sum_{k=1}^{p} a_k \left(\frac{u}{t}\right)^{2k-1} \right) dB_0(u). \tag{5.80$'$}$$

It is not easy to give a simple formula satisfied by the coefficients a_k; $k = 1, 2, \ldots$, however for small p it is easy to calculate them and to see that the formula (5.80$'$) has the same expression as (5.73) which was given by P. Lévy.

(2°) In the case $N = 2p$, let D_N be the same differential operator as in (1°). Then,

$$D_4 D_6 \cdots D_{2p} X_{2p}(t) = X_2(t). \tag{5.81}$$

This relation can be paraphrased in terms of spectral density functions as follows:

$$f_{2p}(\lambda) = \frac{c_{2p} \cdot c_{2p-2} \cdots c_4}{((4p-3)^2 + \lambda^2)((4p-7)^2 + \lambda^2) \cdots (5^2 + \lambda^2)} \cdot f_2(\lambda).$$

Thus, it suffices for us to investigate $f_2(\lambda)$ or $c_2(\lambda)$. We therefore take the covariance function

$$\gamma_2(h) = \cosh h - \frac{1}{2\pi} \int_0^{2\pi} \{2\cosh(2h) - 2\cos\theta\}^{1/2}\, d\theta,$$

and expand the integrand $\{\cdot\}$ using Legendre polynomials. The Fourier inverse transforms of the terms in the expression are summed up to be

$$f_2(\lambda) = \frac{1}{2\pi} \left(\frac{1}{1+\lambda^2} - \sum_{k=0}^{\infty} \frac{b_k}{(4k+3)^2 + \lambda^2} \right),$$

where $b_k = (4k+3)(a_{k+1} - a_k)^2$ with $a_k = (2k-1)!!/(2k)!!$. This expression tells us that $X_2(t)$ can not satisfy the multiple Markov property. However, the expansion of $f_2(\lambda)$ can be approximated by a finite sum, so that $X_2(t)$ may be regarded as a limit $(n \to \infty)$, in a certain sense, of a sequence of n-ple Markov Gaussian processes $X^{(n)}(t)$. Further, $X_{2p}(t)$ is dealt with in the same manner.

As for the innovation of the Brownian motion $\{X(A);\ A \in R^2\}$ itself, we refer to Si Si (1991).

It is now time to summarize what we have discussed so far on the multiple Markov property of the $M_N(t)$-process. In the case N is odd, there is a remarkable law that the degree of the Markov property increases by one as the dimension N increases by two. While for even N, the situation is quite different. Such a difference coming from the geometric structure of the parameter space suggests that we consider the following.

(i) The peculiar relation between the dimension N and the order of the Markov property of the $M_N(t)$-process seems to come from the fact that the original Brownian motion $X_N = \{X_N(A);\ A \in R^N\}$ has a complex probabilistic structure. This observation suggests that we give further consideration to the question of how to define the multiple Markov property for general Gaussian random fields in the multi-dimensional parameter case.

H. P. McKean (1963) presented an interesting study related to this problem, with which many probabilists have been concerned. On the other hand, to generalize the simplest relation between a Brownian motion B and a white noise $\dot{B}$ in the one- dimensional parameter case, we can start from a multi-dimensional parameter white noise and proceed to form a Brownian motion X_N which is represented by a white noise integral, where the representation may or may not be canonical.

(ii) In the case of even $N(=2p)$, the $X_N(t)$ might be defined to be a $(p+1/2)$-ple Markov Gaussian process, however it is hopeless to find a reasonable correspondence between such a (noninteger order) Markov property and an expression of the spectral density function $f_{2p}(\lambda)$ of $X_{2p}(t)$. One might think of a stationary Gaussian process having a spectral density function of the form $\text{const}\cdot(1+\lambda^2)^{-\alpha}$, to be α-ple Markov, where α (>1) is any real number. Thus, the $X_{2p}(t)$-process would serve to explain how to define an α-ple Markov Gaussian process.

(iii) The next question that naturally arises is the question of how to find a way to introduce infinitely multiple Markov Gaussian process. We now recall the formula for $c_{2p+1}(\lambda)$ (for X_{2p+1}), which is of the form

$$\text{const}\cdot(\text{polynomial of degree } p+1)^{-1}$$

as in (5.79), and we let the degree of the polynomial increase, still being of the form $c(\lambda) = c/P(i\lambda)$ with modification for convergence. Then, an example of an infinitely multiple Markov process could be obtained. As a partial answer to this conjecture, we mention below an example given by K. Urbanik (1962).

EXAMPLE 1. Take a constant $c \neq 0$ and an increasing sequence $\{a_k;\ k-1, 2, \dots\}$ of positive numbers and set

$$c(\lambda) = c \Bigg/ \prod_{k=1}^{\infty}\left(1+\frac{i\lambda}{a_k}\right), \tag{5.82}$$

where the following condition is assumed:

$$\limsup_{k\to\infty}(a_k/a_{k+1}) < 1.$$

Then, $|c(\lambda)|^2$ is integrable and satisfies

$$\int_{-\infty}^{\infty} \frac{\log|c(\lambda)|}{1+\lambda^2}\,d\lambda > -\infty.$$

Furthermore, if we denote the Fourier transform of $c(\lambda)$ by $F(u)$, it is easy to prove that

$$X(t) = \int_{-\infty}^{t} F(t-u)\,dB_0(u) \tag{5.83}$$

is the canonical representation. Thus a purely nondeterministic stationary Gaussian process $X = \{X(t);\ t \in R\}$ is obtained.

This $X(t)$ is differentiable as many times as we wish, however it cannot be governed by any stochastic differential equation of the form (5.9) no matter how large N is chosen. However, it would be expected that X enjoys the Markov property in some sense as is seen from what we have discussed so far. We are thus in a position to reconsider the Markov property as well as its generalization to be a notion that describes the nature of dependency when the time parameter varies. We note again that we always restrict our attention to Gaussian processes.

First recall how the simple Markov property is usually defined. Let $X = \{X(t, \omega);\ t \in I\}$ be a Gaussian process on $(\Omega, \mathbf{B}, P)$. (Here use a probability parameter $\omega \in \Omega$.) Let $\mathbf{B}_t(X)$ and $\mathbf{B}^t(X)$ be the σ-fields generated by $\{X(s);\ s \le t\}$ and $\{X(s);\ s \ge t\}$, respectively. For $A \in \mathbf{B}_t(X)$ and $B \in \mathbf{B}^t(X)$, we have

$$\begin{aligned} P(A \cap B | X(t)) &= E(\chi_A \cdot \chi_B | X(t)) = E\{E(\chi_A \cdot \chi_B | \mathbf{B}_t(X)) | X(t)\} \\ &= E\{\chi_A \cdot E(\chi_B | \mathbf{B}_t(X)) | X(t)\} \quad (\text{since } A \in \mathbf{B}_t(X)) \\ &= E\{\chi_A \cdot E(\chi_B | X(t)) | X(t)\} \quad (\text{from the Markov property}) \\ &= E(\chi_A | X(t)) E(\chi_B | X(t)), \end{aligned}$$

where χ_A is the indicator function of a set A. The last equality is obtained by using the fact that $E(\cdot | X(t))$ is $\sigma(X(t))$-measurable. Thus, we can say that the events A and B are conditionally independent when $X(t)$ is given;

$$P(A \cap B | X(t)) = P(A | X(t)) P(B | X(t)), \quad \text{a.e. } P. \tag{5.84}$$

Conversely, if (5.84) is valid for any $A \in \mathbf{B}_t(X)$ and $B \in \mathbf{B}^t(X)$, then we have

$$\begin{aligned} \int_A P(B | \mathbf{B}_t(X)) \, P(d\omega) &= P(A \cap B) = E\{P(A \cap B | X(t)\} \\ &= E\{P(A | X(t)) P(B | X(t))\} = E\{\chi_A P(B | X(t))\} \\ &= \int_A P(B | X(t)) \, P(d\omega). \end{aligned}$$

Namely, we obtain

$$P(B | \mathbf{B}_t(X)) = P(B | X(t)) \quad \text{a.e. } P, \tag{5.85}$$

which shows that X is a simple Markov Gaussian process.

Thus, equality (5.84) is proved to be equivalent to the simple Markov property.

Such a paraphrasing is easily extended to an N-ple Markov Gaussian process in the restricted sense. If $X = \{X(t);\ t \in I\}$ satisfies the N-ple Markov property in the restricted sense, then we can form an N-dimensional-valued simple Markov process $\mathbf{X} = \{\mathbf{X}(t);\ t \in I\}$,

$$\mathbf{X}(t) = (X(t), X'(t), \dots, X^{(N-1)}(t)).$$

The vector-valued versions of (5.84) and (5.85) hold for the process $\mathbf{X}$, where

$$\sigma(\mathbf{X}(t)) = \sigma(X(t), X'(t), \ldots, X^{(N-1)}(t)).$$

Thus, formula (5.84), for example, turns into the following

$$\begin{aligned} &P(A \cap B | X(t), X'(t), \ldots, X^{(N-1)}(t)) \qquad (5.86) \\ &\quad = P(A|X(t), X'(t), \ldots, X^{(N-1)}(t)) \cdot P(B|X(t), X'(t), \ldots, X^{(N-1)}(t)), \\ &\qquad\qquad A \in \mathbf{B}_t(X), \qquad B \in \mathbf{B}^t(X). \end{aligned}$$

REMARK. Related topics are discussed in Y. Okabe (1974).

This change of expression leads us to define a Markov property with $N = \infty$. We set

$$\partial \mathbf{B}_t \; (= \partial \mathbf{B}_t(X)) = \bigcap_{\varepsilon>0} \sigma\{X(s);\; s \in (t-\varepsilon, t+\varepsilon)\}.$$

If a Gaussian process $X = \{X(t, \omega);\; t \in I\}$ satisfies the condition that for any fixed $t \in I$,

$$P(A \cap B | \partial \mathbf{B}_t) = P(A|\partial \mathbf{B}_t) P(B|\partial \mathbf{B}_t), \qquad A \in \mathbf{B}_t(X),\; B \in \mathbf{B}^t(X), \quad (5.87)$$

then we say that X has a *σ-Markov property in the restricted sense.* (The notation σ means "countably infinitely many".) It is easy to prove that if X has the N-ple Markov property in the restricted sense, it has the σ-Markov property in the restricted sense.

With regard to such a property, the result by N. Levinson and H. P. McKean (1964) deserves mention, where the X in question is assumed to be stationary.

THEOREM 5.6. *Suppose a stationary Gaussian process X has canonical representation with kernel $G(t-u)$ and let $c(\lambda)$ be the Fourier inverse transform of G. Then X has the σ-Markov property in the restricted sense if and only if the function $1/c(\lambda)$ is an entire function of infraexponential type.*

The proof of this Theorem requires additional preparation and, so is omitted from this book. The details of the proof are in the paper by Levinson and McKean. Note that the example by Urbanik, introduced in Example 1, has the σ-Markov property in the restricted sense.

REMARK. The σ-Markov property in the restricted sense is, as it were, infinitely many multiple Markov properties in the restricted sense, however it never implies that $X(t)$ is analytic in t. If it were analytic, then it would be a deterministic Gaussian process.

Before closing this section we make another short remark about canonical representations. Having observed the canonical representation of the M_{2p+1}, the kernel of which is a polynomial in u/t, one may ask if there are any noncanonical kernels in a similar form, namely, another polynomial in t/u. Using the time change defined by (5.65), we are given a stationary Gaussian process X_{2p+1} and the question is reduced to that of finding a possible candidate for $c(\lambda)$, which has to be rational function of the form $c(\lambda) = Q(i\lambda)/P(i\lambda)$ with P and Q polynomials, and for which $|c(\lambda)|^2/2\pi$ is equal to spectral density function. The following example serves to illustrate what we are further interested in with regard to the canonical representation.

EXAMPLE 2. For the process $M_7 = \{M_7(t);\ t \geq 0\}$, we can take

$$c(\lambda) = \left(\frac{c}{(11+i\lambda)(7+i\lambda)(3+i\lambda)(1+i\lambda)}\right)\frac{5-i\lambda}{5+i\lambda}$$

to obtain a noncanonical representation of the stationary process X_7. Using (5.65) in the reverse way, we obtain a noncanonical representation of M_7,

$$M_7(t) = \int_0^t \sqrt{10}\left(\frac{3}{5} - \frac{3u}{t} + \frac{5u^2}{t^2} - \frac{3u^3}{t^3} + \frac{2u^5}{5t^5}\right) dB_0(u).$$

As is easily seen from this example, we can find infinitely many kernels which represent the process M_{2p+1} and are polynomials in u/t. Moreover, such noncanonical representations can provide relations between $\mathbf{B}_t(X)$ and $\mathbf{B}_t(B_0)$, problems in prediction theory, backward representations, and so forth.

§6. T-positivity

The last topic in this chapter is the T-positivity of Gaussian processes. This concept was originally introduced in quantum field theory. As is well known, relativistic quantum field theory is based upon Minkowski space, and the construction of such fields presents difficult mathematical problems. If we take, instead of the real time t, the imaginary time it, then the Minkowski space becomes Euclidean space. In the boson case, Euclidean fields are viewed as multidimensional (4-dimensional) parameter stochastic processes, or what is the same as random fields. Thus, the stochastic process theory provides a powerful tool for the constructive field theory.

After obtaining probabilistic results, one must return to real fields. The crucial part, in this direction, must be the question of finding conditions under which Euclidean fields convert into real time fields. To answer this question, we assume that the parameter space is one-dimensional, so that the random fields in question are stochastic processes. In order to apply our theory we will further assume that the processes are mean continuous, stationary, and Gaussian.

There is a well-known sufficient condition, given by E. Nelson and others, for a Euclidean field to be converted into a relativistic quantum field, namely, it is sufficient that the given process has the Markov property together with some additional properties on the reflection. We are now going to give another sufficient condition, the so-called *T-positivity*.

Before the definition can be given, we need to introduce some notation. Let $X = \{X(t);\ t \in R\}$ be a real stationary Gaussian process. The subspace of $L^2(\Omega, \mathbf{B}(X), P)$ involving all $\mathbf{B}^t(X)$ $(= \sigma\{X(s);\ s \geq t\})$-measurable functions is denoted by $L^2(\Omega, \mathbf{B}^t(X), P)$. Let P^t stand for the projection down to this subspace. Define an operator T on $L^2(\Omega, \mathbf{B}(X), P)$ by

$$TX(t) = X(-t), \qquad t \in R, \text{ and } T1 = 1. \tag{5.88}$$

Then, T is unitary and is called a *reflection*.

Now we give a definition by using P^t evaluated at $t = 0$.

DEFINITION 5.5. A stationary Gaussian process X is said to satisfy T-positivity if

$$P^0TP^0 \geq 0 \text{ (the property of a positive definite operator)}. \tag{5.89}$$

First, a necessary condition for X to satisfy T-positivity is derived when the inequality (5.89) is restricted to the subspace $H(X)$, which is the linear span of $\{X(t);\ t \in R\}$. Recall that we always assume $E(X(t)) = 0$. Since the subspace $P^0H(X)$ is spanned by $\{X(t);\ t \geq 0\}$, it is enough to find a condition for the inequality

$$\sum_{j,k=1}^{n} a_i\overline{a}_k E(X(t_j) \cdot TX(t_k)) \geq 0, \tag{5.90}$$

to hold for any n, for any choice of positive t_k and for any choice of complex a_k, $k = 1, 2, \ldots, n$. In terms of the covariance function $\gamma(t)$, (5.90) can be written as

$$\sum_{j,k=1}^{n} a_j\overline{a}_k\gamma(t_j + t_k) \geq 0. \tag{5.91}$$

Noting that $\gamma(t)$ is bounded and continuous, we appeal to Bernstein's theorem (1928) to establish the following representation of $\gamma(t)$:

$$\gamma(t) = \int_0^\infty e^{-tx}\, m(dx), \qquad t \geq 0, \tag{5.92}$$

where m is a bounded Borel measure on $[0, \infty)$. Thus, we have found a necessary condition (5.92) for T-postivity in terms of $\gamma(t)$.

By using the canonical representation of X, we are now able to see more precisely how the operator acts on $H(X)$.

Start with the spectral representation of $\gamma(t)$ of the Gaussian process X in question,

$$\gamma(t) = \int_{-\infty}^{\infty} e^{i\lambda t} f(\lambda)\, d\lambda, \qquad (f(\lambda) = f(-\lambda)).$$

Let the canonical representation of X be given by

$$X(t) = \int_{-\infty}^{t} F_-(t-u)\, dB_-(u), \tag{5.93}$$

and let the "backward" canonical representation exist and be given by

$$X(t) = \int_{t}^{\infty} F_+(t-u)\, dB_+(u). \tag{5.94}$$

The representation (5.94) is the one satisfying the condition

$$\sigma\{X(s);\ s \geq t\} = \sigma\{B_+(s) - B_+(u);\ s, u \geq t\}.$$

Such a representation is constructed in the following manner. First, we recall that in §3.4, a suitable $c(\lambda)$ is chosen so as to satisfy $|c(\lambda)|^2 = 2\pi f(\lambda)$. In addition, $c(w)$ has to be an element of the Hardy class $\mathbf{H}_2$ in the lower half plane ($\mathrm{Re}\, w < 0$). Quite contrary to this case, the backward canonical representation requires that we select $c_+(w)$ from the class $\mathbf{H}_2$ in the upper half plane. The kernel F_+ is then obtained as the Fourier transform of $c_+(\lambda)$, the boundary value.

In order to study how the operator T acts in this situation, it is convenient to use the spectral representation of X,

$$X(t) = \int e^{it\lambda}\, dZ(\lambda), \qquad E(|dZ(\lambda)|^2) = f(\lambda)\, d\lambda. \tag{5.95}$$

The Fourier inverse transforms of F_- and F_+, namely, c_- and c_+ satisfy the relations

$$|c_-(\lambda)|^2 = |c_+(\lambda)|^2 = 2\pi f(\lambda) \quad \text{and} \quad \overline{c_-(\lambda)} = c_+(\lambda).$$

We rewrite (5.95) in the form

$$X(t) = \int e^{it\lambda} c_-(\lambda) \frac{dZ(\lambda)}{c_-(\lambda)} = \int e^{it\lambda} c_+(\lambda) \frac{dZ(\lambda)}{c_+(\lambda)},$$

from which the representations (5.93) and (5.94) are formally derived by Parseval's formula. If the use of the formal notations $\dot{B}_\pm(u)$ and $\dot{Z}(\lambda)$ are permitted, then we have

$$\widehat{\dot{B}}_-(\lambda) = \overline{\dot{Z}(\lambda)}/c_+(\lambda), \qquad \widehat{\dot{B}}_+(\lambda) = \overline{\dot{Z}(\lambda)}/c_-(\lambda),$$

where the notation ^ denotes the Fourier transform. With this background, we can now apply T to $X(t)$:

$$TX(t) = \int e^{-it\lambda} \dot{Z}(\lambda)\, d\lambda = \int e^{it\mu}\, \overline{\dot{Z}(\mu)}\, d\mu.$$

Hence we obtain $T\dot{Z}(\lambda) = \overline{\dot{Z}(\lambda)}$, which implies

$$T(\overline{\dot{Z}(\lambda)}/c_+(\lambda)) = \dot{Z}(\lambda)/c_-(\lambda) = \overline{(\overline{\dot{Z}(\lambda)}/c_-(\lambda))}.$$

This means that $T\widehat{\dot{B}}_-(\lambda) = \widehat{\dot{B}}_+(\lambda)$. Finally, we have

$$T\dot{B}_+(u) = \dot{B}_-(-u). \tag{5.96}$$

Another interesting result deserves to be mentioned. Namely,

Theorem 5.7. *For φ, $\psi \in L^2(R)$, we have*

$$E\left(\int \varphi(u)\, dB_-(u) \int \psi(u)\, dB_+(u)\right) = \int \overline{\hat{\varphi}(\lambda)}\hat{\psi}(\lambda)(c_-(\lambda)/c_+(\lambda))\, d\lambda. \tag{5.97}$$

Proof. The equality (5.97) follows from a calculation below. The left-hand side of (5.97) is equal to

$$E\left\{\iint \overline{(\dot{Z}(\lambda)/c_+(\lambda))\hat{\varphi}(\lambda)}\cdot(\dot{Z}(\lambda')/\overline{c_-(\lambda')})\hat{\psi}(\lambda')\,d\lambda\,d\lambda'\right\}$$

$$= \iint \frac{\overline{\varphi(\lambda)}\hat{\psi}(\lambda')}{\overline{c_+(\lambda)c_-(\lambda')}}|f(\lambda)|^2\delta_{\lambda-\lambda'}\,d\lambda\,d\lambda'$$

$$= \int \overline{\varphi(\lambda)}\hat{\psi}(\lambda)\frac{c_-(\lambda)}{c_+(\lambda)}\,d\lambda. \qquad \square$$

Remark. Using this Theorem and the relation

$$P_0\int \psi(u)\,dB_+(u) = \int \psi(u)\chi_{[0,\infty)}(u)\,dB_+(u),$$

we can obtain a necessary condition for T-positivity to be equivalent to the inequality (5.91).

Next we study the condition for T-positivity on the whole space $L^2(\Omega, \mathbf{B}(X), P)$. The subspace consisting of $\mathbf{B}^0(X)$-measurable functions in $L^2(\Omega, \mathbf{B}(X), P)$ is spanned by the algebra of functions of the form $\exp[i\sum_k a_k X(t_k)]$, $t_k \geq 0$, $a_k \in R$. Choosing the t_k as in (5.90), condition (5.89) is equivalent to the fact that the matrix

$$E\left\{\exp\left[i\sum_k a_k X(t_k)\right]\cdot\exp\left[-i\sum_k a_k' X(-t_k)\right]\right\} \tag{5.98}$$

is positive definite. There we have used the fact that T acts multiplicatively,

$$T\exp\left[\sum_k a_k' X(t_k)\right] = \exp\left[\sum_k a_k' TX(t_k)\right].$$

The matrix (5.98) is easily calculated and we can see that it is equal to

$$\exp\left[-\frac{1}{2}\sum_{k,j}((a_k a_j + a_k' a_j')\gamma(t_k - t_j) + 2a_k a_j'\gamma(t_k + t_j))\right]$$

and that the form inside of [ ] is conditionally positive definite. Therefore, (5.92) is also a sufficient condition for (5.98) to be positive definite on the space $L^2(\Omega, \mathbf{B}(X), P)$. Summing up:

Theorem 5.8. *A mean continuous stationary Gaussian process X satisfies T-positivity if and only if the covariance function $\gamma(t)$ has the form in* (5.92).

Example. Motivated by quantum field theory, we consider a double Markov stationary Gaussian process X_a with function $c_-(\lambda)$ of the form

$$c_-(\lambda) = \frac{1}{m+i\lambda} + a\frac{1}{M+i\lambda}, \qquad m, M \geq 0, \ a \in R.$$

(The process under consideration corresponds to two free particles with mass m and M, respectively, in physical terminology.) Let a vary. Then it is interesting to find the range of a for which X_a satisfies T-positivity. Our result says that X_a for $a \in (0, \infty)$ satisfies T-positivity if $M \neq m$. The point $a = 0$ is "critical", since X_0 is a simple Markov process.

We recommend that the reader see the papers by Y. Okabe (1979) and (1981) for additional interesting results on Gaussian processes with T-positivity.

CHAPTER VI

Equivalence of Gaussian Processes

§1. Setting up the problem

A stochastic process $X = \{X(t);\ t \in I\}$ induces a probability measure μ on R^I. This was discussed briefly in §1.6. More precisely, for a stochastic process X defined on a probability space $(\Omega, \mathbf{B}, P)$, we introduce a probability measure μ on a measurable space (R^I, Σ^I) in such a way that

$$\mu\{x \in R^I;\ (x(t_1), \dots, x(t_n)) \in B)\} = P((X(t_1, \omega), \dots, X(t_n, \omega)) \in B)),$$

for any $t_1, \dots, t_n$ and for any n-dimensional Borel subset B of R^n. For example, if X is a Brownian motion $B = \{B(t);\ t \in [0, T]\}$, then the induced measure μ, called a *Wiener measure*, can be regarded as a measure supported by the space $C = C([0, T])$ consisting of continuous functions on $[0, T]$.

Let two stochastic processes $X_i = \{X_i(t);\ t \in I\}$, $i = 1, 2$, be given an let their induced measures on (R^I, Σ^I) be μ_i, $i = 1, 2$, respectively. We denote the two measure spaces (R^I, Σ^I, μ_i), $i = 1, 2$, by $(\Omega, \mathbf{B}, P_i)$, $i = 1, 2$, respectively (cf. Theorem 1.13). With this notation we give

DEFINITION 6.1. Let P_1 and P_2 be the induced measures from stochastic processes X_1 and X_2, respectively. Then X_1 is said to be *absolutely continuous* with respect to the process X_2, if P_1 is absolutely continuous with respect to P_2. If in addition X_2 is absolutely continuous with respect to X_1, then they are called *equivalent*. While, if there exists an event $B \in \mathbf{B}$ such that $P_1(B) = 1$ and $P_2(B) = 0$, then X_1 and X_2 are called *singular*.

As is easily seen, if X_1 and X_2 are equivalent, they share the events of probability one (or zero) with respect to the probability measures P_1 and P_2; that is, for $A \in \mathbf{B}$, $P_1(A) = 1$ (or 0) if and only if $P_2(A) = 1$ (or 0). This means that equivalence of processes tells us that two stochastic processes have the same probabilistic properties which hold with probability one (or 0). Also, as we will see later, in the case of equivalence, the density $\varphi(\omega) = P_2(d\omega)/P_1(d\omega)$ plays a very important role in our approach. Next we give a definition of Gaussian measure. Let the measurable space (R^I, Σ^I) be provided as before.

DEFINITION 6.2. A probability measure P on (R^I, Σ^I) is said to be

Gaussian, if it is induced by a Gaussian process $X = \{X(t);\ t \in I\}$.

If, in particular, the measure P is supported by the space $\mathbf{C} = \mathbf{C}(I)$ of continuous functions on I, then P is called a Gaussian measure on $\mathbf{C}$ (or, more precisely, on $(\mathbf{C}, \mathbf{B})$, $\mathbf{B}$ being the topological Borel field).

We often meet examples of such a measure.

REMARK. There will be no difficulty to consider more general Gaussian measure on $(R^\Lambda, \Sigma^\Lambda)$ induced by a Gaussian system $X = \{X_\lambda;\ \lambda \in \Lambda\}$. There Σ^Λ is the σ-field generated by cylinder subsets of Λ.

§2. General theory of equivalence of Gaussian measures

We will first state some general properties of Gaussian measures, in particular on infinite dimensional vector spaces, as background for our discussion of the equivalence problem for Gaussian measures. A criterion for equivalence is presented in terms of entropy, and by using this criterion, we will prove the surprising dichotomy that two Gaussian measures are *either* equivalent or singular. Such a dichotomy comes essentially from the theorem of Kakutani (S. Kakutani, 1948) that deals with infinite products of measures.

DEFINITION 6.3. Let P_i, $i = 1, 2$, be probability measures on a measurable space $(\Omega, \mathbf{B})$ and let $\alpha = \{A \in \mathbf{B};\ i = 1, 2, \dots, n\}$ be a finite partition of Ω. Then the quantity

$$H(P_2|P_1) = \sup_\alpha \sum_{i=1}^n P_2(A_i) \log \frac{P_2(A_i)}{P_1(A_i)} \tag{6.1}$$

is called the *entropy* of the probability measure P_2 relative to P_1, where α runs through the class of all possible finite partitions of Ω. Note that, in the formula for $H(P_2|P_1)$, we tacitly assume that $0 \log 0 = 0$ and $\log 0 = -\infty$.

Before we can prove the theorem we need several lemmas.

LEMMA 6.1. (i) *If a probability measure P_2 is absolutely continuous with respect to another probability measure P_1, then the entropy $H(P_2|P_1)$ can be expressed, by using the Radon-Nikodym density $\varphi(\omega) = P_2(d\omega)/P_1(d\omega)$, in the form*

$$H(P_2|P_1) = \int_\Omega \log \varphi(\omega)\, P_2(d\omega) = \int_\Omega (\log \varphi(\omega))\varphi(\omega)\, P_1(d\omega). \tag{6.2}$$

(ii) *While, if P_2 is not absolutely continuous with respect to P_1, then $H(P_2|P_1) = \infty$.*

PROOF. Statement (ii) is easily seen from the formula (6.1). The assertion (i) is proved by using a Jensen type inequality for a probability measure Φ on $[0, \infty)$:

$$\int_0^\infty x \log x\, \Phi(dx) \geq \left(\int_0^\infty x\, \Phi(dx)\right) \left(\log \int_0^\infty x\, \Phi(dx)\right). \tag{6.3}$$

The inequality (6.3) can be derived from the convexity of $x \log x$. Now for any event $A \in \mathbf{B}$ with $P_1(A) > 0$, we consider a distribution function

$F_A(x) = \Phi_A([0, x]) = P_1(A \cap \{\varphi(\omega) \le x\})/P_1(A)$. Obviously

$$\int_A \{\log \varphi(\omega)\}\varphi(\omega)P_1(\omega) \ge P_2(A)\log\{P_2(A)/P_1(A)\}$$

holds. Hence, for any finite partition $\alpha = \{A_1, A_2, \dots, A_n\}$, the inequality

$$\int_\Omega \{\log \varphi(\omega)\}\, P_2(d\omega) \ge \sum_{i=1}^{n} P_2(A_i)\log\{P_2(A_i)/P_1(A_i)\}$$

must hold. Thus, $\int_\Omega \{\log \varphi(\omega)\}\, P_2(d\omega) \ge H(P_2|P_1)$ is proved.

We next prove the opposite inequality. We construct a finite partition α_n of Ω by using the values of $\varphi(\omega) \ge 0$:

$$\alpha_n = \{A_{n,k};\, k = 1, 2, \dots, n^2 + 1\},$$

where $A_{n,k} = \{\omega;\, (k-1)/n \le \varphi(\omega) < k/n\}$, $k = 1, 2, \dots, n^2$, and $A_{n,n^2+1} = \{\omega;\ \varphi(\omega) \ge n\}$. Set

$$\varphi_n(\omega) = \frac{k-1}{n} \quad \text{for } \omega \in A_{n,k},\ k = 1, 2, \dots, n^2 + 1.$$

Then we have

$$\begin{aligned}\sum_{k=1}^{n^2+1} P_2(A_{n,k})\log\frac{P_2(A_{n,k})}{P_1(A_{n,k})} &\ge \sum_{k=1}^{n^2+1}\int_{A_{n,k}} \varphi_n(\omega)\log\varphi_n(\omega)P_1(d\omega)\\ &\ge \int_\Omega \varphi_n(\omega)\log\varphi_n(\omega)\, P_1(d\omega).\end{aligned}$$

Using the facts that $x\log x$ is bounded below and that $\varphi_n(\omega)\log\varphi_n(\omega)$ converges to $\varphi(\omega)\log\varphi(\omega)$ monotonically as $n \to \infty$, we see that the right-hand of the inequalities converges to $\int_\Omega \varphi(\omega)\log\varphi(\omega)\, P_1(d\omega)$. Thus, (6.2) is proved. □

LEMMA 6.2. *Let* $\mathbf{B}^\sim$ *be a sub-σ-algebra of* $\mathbf{B}$ $(\mathbf{B}^\sim \subset \mathbf{B})$, *and let* $P_1^\sim$ *and* $P_2^\sim$ *be the restrictions of* P_1 *and* P_2 *to* $\mathbf{B}^\sim$, *respectively,* $P_i^\sim(A) = P_i(A)$ *for* $A \in \mathbf{B}^\sim$, $i = 1, 2$. *Then* $H(P_2^\sim|P_1^\sim) \le H(P_2|P_1)$.

PROOF. Since a finite partition of Ω with members in $\mathbf{B}^\sim$ is the one with those in $\mathbf{B}$, the inequality follows immediately from Definition 6.1 of H. □

LEMMA 6.3. *Suppose that an increasing sequence* $\mathbf{B}_n$, $n = 1, 2, \dots$ *of sub-σ-fields converges to* $\mathbf{B}$: $\mathbf{B} = \bigvee_n \mathbf{B}_n$. *Let* P_1^n *and* P_2^n *be the restrictions of the given probability measures* P_1 *and* P_2 *to the* $\mathbf{B}_n$, *respectively. If* $\sup_n H(P_2^n|P_1^n) < \infty$, *then* P_2 *is absolutely continuous with respect to* P_1 *and*

$$H(P_2|P_1) = \sup_n H(P_2^n|P_1^n). \tag{6.4}$$

PROOF. From Lemma 6.1 (ii) we see that P_2^n is absolutely continuous with respect to P_1^n. The density $(dP_2/dP_1)(\omega)$ is denoted by $\varphi_n(\omega)$. By assumption we have

$$H(P_2^n|P_1^n) = \int_\Omega \log \varphi_n(\omega)\, P_2^n(d\omega)$$
$$= \int_\Omega (\log \varphi_n(\omega))\varphi_n(\omega)\, P_1^n(d\omega) \le M < \infty \quad \text{for some } M > 0. \tag{6.5}$$

As is easily seen the system $\{\varphi_n(\omega), \mathbf{B}_n;\ n = 1, 2, \dots\}$ forms a martingale with respect to the measure P_1, and $\lim_{n\to\infty} \int_\Omega \varphi_n(\omega)\, P_1(d\omega) < \infty$ follows from (6.5). We can therefore apply Theorem 1.11, to prove that there exists the limit $\varphi_\infty(\omega) = \lim_{n\to\infty} \varphi_n(\omega)$, a.e. P_1. Hence, it can be shown that the extended system $\{\varphi_n(\omega), \mathbf{B}_n;\ n = 1, 2, \dots, \infty\}$ ($n = \infty$ is included) is again a martingale, where $\mathbf{B}_\infty$ may be taken to be the basic σ-field $\mathbf{B}$. To prove the assertion, it suffices to show that $E_1[\varphi_\infty] = \int_\Omega \varphi_\infty(\omega)\, P_1(d\omega) = 1$. The reason is as follows. We have the inequality $E_1[\varphi_\infty|\mathbf{B}_n] \le \varphi_n(\omega)$ (a.s. P_1) for every n (this follows from Fatou's theorem). But if there were some n such that $\int_B \varphi_\infty(\omega) P_1(d\omega) < \int_B \varphi_n(\omega)\, P_1(d\omega)$ on $B \in \mathbf{B}$ with $P_1(B) > 0$, then

$$E_1[\varphi_\infty] = E_1[E_1[\varphi_\infty|\mathbf{B}_n]] < E_1[\varphi_n] = 1$$

must hold. Therefore, the equality $E_1[\varphi_\infty] = 1$ asserts that no such B can exist. That is, the extended system is a martingale.

Now we return to the proof of $E_1[\varphi_\infty] = 1$. Using Tchebyshev's inequality, we have

$$\sup_n \int_{\{\omega;\, \varphi_n(\omega) > K\}} \varphi_n(\omega)\, P_1(d\omega) \le \sup_n E_1[\varphi_n(1 \vee \log \varphi_n)]/\log K \le M_0/\log K$$

for some $M_0 > 0$. Thus, for any $\varepsilon > 0$, there exists a $K > 0$ such that

$$E_1(|\varphi_n| \wedge K) > 1 - \varepsilon, \qquad n = 1, 2, \dots, \tag{6.6}$$

and so we obtain $E_1[\varphi_\infty \wedge K] \ge 1 - \varepsilon$ from the dominated convergence theorem. Therefore $E_1[\varphi_\infty] = 1$, since ε (> 0) is arbitrary. The fact that $\{\varphi_n, \mathbf{B}_n;\ n = 1, 2, \dots, \infty\}$ is a martingale, asserts that for $A \in \mathbf{B}_n$, $n = 1, 2, \dots$, the following equalities

$$P_2(A) = P_2^n(A) = \int_A \varphi_n(\omega) P_1(d\omega) = \int_A \varphi_n(\omega)\, P_1(d\omega) = \int_A \varphi_\infty(\omega)\, P_1(d\omega)$$

hold. In particular, we have $P_2(A) = \int_A \varphi(\omega)\, P_1(d\omega)$ for any $A \in \mathbf{B} = \bigvee_n \mathbf{B}_n$.

As for the equality (6.4), "$\ge$" is trivial, and "$\le$" is easily obtained from Fatou's theorem. □

Remark. For a system $\{\varphi_n(\omega);\ n = 1, 2, \dots\}$ of random variables (not necessarily a martingale), if $E[|\varphi_n|]$ is bounded and if

$$\sup_n \int_{\{\omega;\, |\varphi_n(\omega)| > K\}} \varphi_n(\omega)\, P(d\omega) \to 0,$$

as $K \to \infty$, then the system is called *uniformly integrable* (or *regular*). If, in addition, the limit $\varphi(\omega) = \lim_{n\to\infty} \varphi_n(\omega)$ exists, then uniform integrability is equivalent to $E[|\varphi - \varphi_n|] \to 0$ $(n \to \infty)$. Note that the proof of Lemma 6.3 also shows that uniform integrability of the martingales $\{\varphi_n(\omega), \mathbf{B}_n; n = 1, 2, \dots, \}$ comes from the assumption $\sup_n H(P_2^n|P_1^n) < \infty$.

Lemmas 6.1–6.3, lead to an explicit expression for the entropy of finite dimensional Gaussian measures.

LEMMA 6.4. *Let $\{X_k(\omega);\ k = 1, 2, \dots, n\}$ be a system of random variables on a measurable space $(\Omega, \mathbf{B})$. Assume that the X_k are independent and Gaussian in distribution with respect to both probability measures P_1 and P_2, and assume*

$$E_1[X_k] = m_k, \quad V_1[X_k] = \sigma_k^2 > 0, \quad E_2[X_k] = 0, \quad V_2[X_k] = 1, \quad k = 1, 2, \dots, n,$$

where E_i and V_i are the mean and variance, respectively, with respect to P_i $(i = 1, 2)$. Let $\mathbf{B}'$ be the sub-σ-algebra generated by the system $\{X_i;\ i = 1, 2, \dots, n\}$. Then, the restricted measures P_1' and P_2' of P_1 and P_2, respectively, to $\mathbf{B}'$ are equivalent and the following relations hold:

$$\log\{P_2'(d\omega)/P_1'(d\omega)\} = \log \varphi'(\omega) = \frac{1}{2}\sum_{k=1}^{n}\left\{\log \sigma_k^2 + \frac{(X_k - m_k)^2}{\sigma_k^2} - X_k^2\right\},$$

$$E_1(\log \varphi') = -\frac{1}{2}\sum_{k=1}^{n}\{\sigma_k^2 - 1 - \log \sigma_k^2 + m_k^2\},$$

$$E_2(\log \varphi') = \frac{1}{2}\sum_{k=1}^{n}\{1/\sigma_k^2 - 1 + \log \sigma_k^2 + m_k^2/\sigma_k^2\},$$

$$V_1(\log \varphi') = \frac{1}{2}\sum_{k=1}^{n}\{(\sigma_k^2 - 1)^2 + 2m_k^2\sigma_k^2\},$$

and

$$V_2(\log \varphi') = \frac{1}{4}\sum_{k=1}^{n}\{(\sigma_k^2 - 1)^2/\sigma_k^4 + 2m_k^2/\sigma_k^4\}.$$

Proofs of these results are easily given by direct calculations of the finite dimensional Gaussian densities.

Next we consider Gaussian systems of infinitely many variables, in particular, Gaussian processes. Suppose a system of random variables $X = \{X(t);\ t \in I\}$ with continuous parameter has been given and assume that X is a Gaussian processes with respect to both probability measures P_1 and P_2. To be specific, we take the basic measurable space $(\Omega, \mathbf{B})$ to be (R^I, Σ^I), on which two Gaussian measures P_1 and P_2 are given. Let $\mathbf{B}'$ be the sub-σ-algebra of $\mathbf{B}$ generated by the $X(t)$, $t \in I'$, I' a finite subset of I.

Definition 6.4. The *entropy distance* $\mathbf{I}$ between two measures P_1 and P_2 are defined by

$$\mathbf{I} = \sup_{I' \in \Delta} \mathbf{I}(P_1'|P_2'), \qquad \mathbf{I}(P_1'|P_2') \equiv H(P_1'|P_2') + H(P_2'|P_1'), \tag{6.7}$$

where Δ is the class of all finite subsets I' of I, and where P_1' and P_2' are the restrictions of P_1 and P_2 to $\mathbf{B}'$, respectively.

Using the entropy distance, we obtain the following results.

Lemma 6.5. *Let two measures P_1 and P_2 on $(\Omega, \mathbf{B})$ be Gaussian. If the entropy distance $\mathbf{I}$ is infinite, then P_1 and P_2 are singular.*

Proof. We may assume that $E_2[X(t)] = 0$, $t \in I$. If not, we can take $X^{\sim}(t) = X(t) - E_2[X(t)]$ in place of $X(t)$. Since $\sigma\{X(t);\ t \in I\} = \sigma\{X^{\sim}(t);\ t \in I\}$, our discussion is unchanged. For any finite subset $I' = \{t_1, t_2, \ldots, t_n\}$, the covariance matrices of $\{X(t_k);\ k = 1, 2, \ldots, n\}$ with respect to P_1 and P_2 are denoted by

$$\mathbf{V}_1 = (V_1(t_i, t_j))_{i,j=1,2,\ldots,n} \quad \text{and} \quad \mathbf{V}_2 = (V_2(t_i, t_j))_{i,j=1,2,\ldots,n},$$

respectively. Then there exists a normal matrix $\mathbf{C} = (C_{ij})$ of order n such that $\mathbf{C}^*\mathbf{V}_1\mathbf{C}$ and $\mathbf{C}^*\mathbf{V}_2\mathbf{C}$ are simultaneously diagonalized and one of them, say $\mathbf{C}^*\mathbf{V}_2\mathbf{C}$, is the unit matrix. Set

$$Y_k = \sum_{j=1}^{n} C_{kj} X(t_j), \qquad k = 1, 2, \ldots, n,$$

then, $\mathbf{Y} = \{Y_k;\ k = 1, 2, \ldots, n\}$ is an independent Gaussian system with respect to both probability measures P_1 and P_2 with

$$E_2[Y_k Y_j] = \delta_{kj} \quad \text{(the Kronecker delta)}.$$

Set $\mathbf{B}' = \sigma(\mathbf{Y})$. Then $\sigma(\mathbf{Y}) = \sigma\{X(t_1), \ldots, X(t_n)\}$. Using Lemma 6.4 we have

$$\mathbf{I}(P_1'|P_2') = \frac{1}{2}\sum_{k=1}^{n}\left\{\frac{(\sigma_k^2 - 1)^2}{\sigma_k^2} + m_k^2\left(1 + \frac{1}{\sigma_k^2}\right)\right\}, \tag{6.8}$$

where $m_k = E_1[Y_k]$, $\sigma_k^2 = V_1[Y_k]$ and where P_1' and P_2' are the restrictions of P_1 and P_2 to $\mathbf{B}'$, respectively. Our proof of $\mathbf{I}(P_1|P_2) = \infty$ is divided into two cases.

The first case assumes that there exist two positive constants c_1 and c_2 satisfying

$$c_1 \le \sigma_k \le c_2, \qquad k = 1, 2, \ldots, n, \tag{6.9}$$

for any finite subset $I' = \{t_1, t_2, \ldots, t_n\}$ of I. By Lemma 6.4, there are positive constants a and b such that

$$a\sum_{k=1}^{n}\{(\sigma_k^2-1)^2+m_k^2\}\le \mathbf{I}(P_1'|P_2')\le b\sum_{k=1}^{n}\{(\sigma_k^2-1)^2+m_k^2\},$$

$$V_1[\log\varphi']\le b\sum_{k=1}^{n}\{(\sigma_k^2-1)^2+m_k^2\} \quad \text{and} \quad V_2[\log\varphi']\le b\sum_{k=1}^{n}\{(\sigma_k^2-1)^2+m_k^2\},$$

hold, where $\varphi'(\omega)=P_2'(d\omega)/P_1'(d\omega)$. Choose an event A' with

$$A'=\left\{\omega;\ -\log\varphi'(\omega)>\frac{1}{2}\mathbf{I}(P_1'|P_2')-E_2[\log\varphi']\right\} \tag{6.10}$$

in $\mathbf{B}'$. Then the Tchebyshev inequality guarantees the following relations with c a positive constant:

$$P_2(A')=P_2'(A')\le V_2[\log\varphi']\left\{\frac{1}{2}\mathbf{I}(P_1'|P_2')\right\}^{-2}\le c\{I(P_1'|P_2')\}^{-1},$$

and

$$\begin{aligned}P_1(A')=P_1'(A')\ge 1-P_1'(\Omega-A')&\ge 1-V_1[\log\varphi']\left\{\frac{1}{2}I(P_1'|P_2')\right\}^{-2}\\ &\ge 1-c\{\mathbf{I}(P_1'|P_2')\}^{-1}.\end{aligned}$$

From the assertion $\mathbf{I}=\infty$, there exist, for any $\varepsilon>0$, a finite subset A_ε' of I and an event $A_\varepsilon'=A'$, given by (6.10), such that

$$P_1(A_\varepsilon')\ge 1-\varepsilon \quad \text{and} \quad P_2(A_\varepsilon')\le\varepsilon.$$

This implies that P_1 and P_2 are singular.

The second case is concerned with the situation where at least one of the positive constants c_1 and c_2 in (6.9) is not finite. Then for any $\varepsilon>0$, there exist both a finite subset $T'=\{t_1,t_2,\dots,t_n\}$ of T and a $k\in\{1,2,\dots,n\}$ such that $\sigma_k<\varepsilon$ or $\sigma_k>1/\varepsilon$. We consider the latter case. For the event

$$A_\varepsilon'=\{\omega;\ |Y_k(\omega)-m_k|>\sqrt{\sigma_k}\}$$

in $\mathbf{B}'\subset\mathbf{B}$, we obtain the evaluations

$$\begin{aligned}P_1(A_\varepsilon')=P_1'(A_\varepsilon')&=\frac{1}{\sqrt{2\pi}\sigma_k}\int_{|y-m_k|<\sqrt{\varepsilon}}e^{-(y-m_k)^2/2\sigma_k^2}\,dy\\ &=\frac{1}{\sqrt{2\pi}}\int_{|x|>\sqrt{1/\sigma_k}}e^{-x^2/2}\,dx\ge\frac{1}{\sqrt{2\pi}}\int_{|x|>\varepsilon}e^{-x^2/2}\,dx\end{aligned}$$

and

$$P_2(A_\varepsilon')\le\frac{1}{\sqrt{2\pi}}\int_{|x-m_k|>\sqrt{1/\varepsilon}}e^{-x^2/2}\,dx.$$

Letting $\varepsilon\to 0$, we see that $P_1(A_\varepsilon')\to 1$ and $P_2(A_\varepsilon')\to 0$. This means that P_1 and P_2 are singular. □

The results above imply the famous dischotomy stated in the following theorem.

THEOREM 6.1. *Two Gaussian measures on a measurable space* $(\Omega, \mathbf{B})$ *are either equivalent or singular.* (*No intermediate state is permitted.*)

PROOF. If $\mathbf{I} < \infty$, then $H(P_1|P_2\mathbf{I} < \infty$ and $H(P_2|P_1) < \infty$ hold. By using Lemma 6.3, we can see that the two measures are equivalent. If $\mathbf{I} = \infty$, then the assertion is Lemma 6.5. □

§3. Equivalence of Gaussian processes with discrete parameter

The results established in the previous section are used in this section to study the equivalence problem for discrete parameter Gaussian processes. The basic probability space $(\Omega, \mathbf{B})$ is taken to be (R^{I^+}, Σ^{I^+}), $I^+ = \{1, 2, \dots\}$, on which two Gaussian measures P_1 and P_2 are given. We consider the equivalence of P_1 and P_2, with emphasis on the role of canonical representations of the Gaussian processes X_i which determine these two Gaussian measures P_i, $i = 1, 2$, respectively.

We assume in this section that each $X_i = \{X_{i,n};\ n \in I^+\}$ has a coordinate representation such that $X_{i,n}(x) = X_n(x) = x_n$, $i = 1, 2$, for $x = (x_1, x_2, \dots, x_n, \dots) \in \Omega$, and assume that the sequence (process) $X = \{X_n(x)\}$ is a Gaussian system with respect to both of the measures P_1 and P_2. In addition, suppose that with respect to P_2 the system X is independent identically $N(0, 1)$ distributed; such a system as X is called a *standard independent* (*Gaussian*) *sequence.* Our main purpose is to find conditions under which P_1 and P_2 are equivalent. Namely, we want to determine the class of all Gaussian systems which are equivalent to the standard independent sequence.

First, we consider the case where $E_1[X_i] = 0$, $i \in I^+$. We now recall the canonical representation theory for discrete parameter Gaussian processes (sequence) discussed in §2.4. The canonical representation of a Gaussian process X on the probability space $(\Omega, \mathbf{B}, P_1)$ is obtained by Schmidt's orthogonalization method:

$$X_i = \sum_{j \le i} a_{ij}\xi_j, \qquad i \in I^+, \tag{6.11}$$

where the system $\Xi = \{\xi_i,\ i \in I^+\}$ is a standard independent sequence, in other words, Ξ is the innovation process for X in $(\Omega, \mathbf{B}, P_1)$. We may assume $a_{ii} > 0$. Otherwise $(a_{ii} = 0)$, the X_i can be omitted from the system X. For the later purpose (6.11) is rewritten as

$$X_i = (1 - \gamma_i)\xi_i + \sum_{j<i} \xi_j, \qquad 1 - \gamma_i = a_{ii}. \tag{6.12}$$

Assume now that two Gaussian measures P_1 and P_2 on $(\Omega, \mathbf{B})$ are equivalent. Denote the Radon-Nikodym density $P_2(d\omega)/P_1(d\omega)$ by $\varphi(\omega)$ and define a martingale $M = \{M_n, \mathbf{B}_n(x);\ n = 1, 2, \dots\}$ with respect to P_2 where M_n and $\mathbf{B}_n(X)$ are given by

$$M_n = E_2[1/\varphi|\mathbf{B}_n(X)] \quad \text{and} \quad \mathbf{B}_n(X) = \sigma\{X_i;\ i = 1, 2, \dots, n\}.$$

An explicit expression for M_n is obtained in the following manner. For any $n \in I^+$, we set $P_1^{(n)}(d\omega) = M_n(\omega)P_2(d\omega)$. Then $P_1^{(n)}$ is viewed as the restriction of P_1 to the sub-σ-field $\mathbf{B}_n(X)$. Hence, M_n is determined by the Radon-Nikodym density obtained by the two finite dimensional Gaussian measures whose covariances are given by

$$\begin{aligned} (V_1^{(n)}(i, j)) &= (V_1(i, j)) = \left(\sum_{k=1}^{i\wedge j} a_{ik}a_{jk}\right), \\ (V_2^{(n)}(i, j)) &= (\delta_{ij}). \end{aligned} \tag{6.13}$$

Now recall the assumption that $a_{ii} = 1 - \gamma_i > 0$, $i = 1, 2, \dots, n$. The linear equation (6.12) for ξ_i, $i = 1, 2, \dots, n$, is solved in the form

$$\xi_i = \frac{\left\{X_i - \sum_{j=1}^{i-1} a_{ij}\xi_j\right\}}{1-\gamma_i} = \frac{\left\{X_i - \sum_{j=1}^{i-1} b_{ij}X_j\right\}}{1-\gamma_i}, \tag{6.14}$$

where the coefficients b_{ij}, $i, j = 1, 2, \dots,$ are independent of n. Since the ratio of densities of two Gaussian distributions on R^n with covariances given by (6.13) is expressed in the form

$$M(x_1, \dots, x_n) = \prod_{i=1}^n \frac{1}{1-\gamma_i} \exp\left[-\frac{1}{2}\left\{\frac{1}{(1-\gamma_i)^2}\left(x_i - \sum_{j=1}^{i-1} b_{ij}x_j\right)^2 - x_i^2\right\}\right],$$
$$(x_1, \dots, x_n) \in R^n.$$

We have therefore

$$\begin{aligned} M_n(\omega) &= \prod_{i=1}^n \frac{1}{1-\gamma_i} \exp\left[-\frac{1}{2}\left\{\frac{1}{(1-\gamma_i)^2}\left(X_i(\omega) - \sum_{j=1}^{i-1} b_{ij}X_j(\omega)\right)^2 - X_i(\omega)^2\right\}\right] \\ &= \prod_{i=1}^n \frac{1}{1-\gamma_i} \exp\left[-\frac{1}{2}\left\{\xi_i^2 - (1-\gamma_i)^2\left(\xi_i - \sum_{j=1}^{i-1} a_{ij}\frac{\xi_j}{1-\gamma_j}\right)^2\right\}\right]. \end{aligned} \tag{6.15}$$

Since $\mathbf{I}$ is finite by Theorem 6.1, we obtain

$$E_1^{(n)}[\log M_n] = E_2[M_n \log M_n] \le E_2[-(\log\varphi)/\varphi] = E_1[-\log\varphi] < \infty.$$

Since the n-dimensional Gaussian random vector $(\xi_1, \xi_2, \dots, \xi_n)$ has covariance $V_1^{(n)}(i, j) = \delta_{i,j}$ with respect to the probability measure $P_1^{(n)}$ (or P_1), the left-hand side of the above inequality is

$$E_1^{(n)}[\log M_n] = \sum_{i=1}^n \left[-\log(1-\gamma_i) - \frac{1}{2}\left\{1 - (1-\gamma_i)^2\left(1 + \sum_{j=1}^{i-1} a_{ij}^2(1-\gamma_j)^{-2}\right)\right\}\right].$$

Noting that

$$-\log(1-\gamma_i) = \gamma_i + \frac{1}{2}\gamma_i^2 + O(\gamma_i^3),$$

we obtain a necessary and sufficient condition for $E_1^{(n)}[\log M_n]$ to be bounded in n. In fact, this condition can be expressed as

$$\sum_{i=1}^{\infty} \gamma_i^2 < \infty \quad \text{and} \quad \sum_{i=1}^{\infty}\sum_{j=1}^{i-1} a_{ij}^2 < \infty .$$

We have proved the following theorem.

THEOREM 6.2. *Let $X = \{X_i\,;\, i \in I^+\}$ be a centered Gaussian process (sequence) defined on a probability space $(\mathbf{\Omega}, \mathbf{B}, P_1)$. Then X is equivalent to the standard independent sequence if and only if the canonical representation of X has the form*

$$X_i = (1-\gamma_i)\xi_i + \sum_{j=1}^{i-1} a_{ij}\xi_j, \qquad i \in I^+, \tag{6.16}$$

in terms of the innovation process $\{\xi_i\,;\, i \in I^+\}$, where

$$\gamma_i < 1 \quad \textit{for all } i \in I^+; \qquad \sum_{i=1}^{\infty}\gamma_i^2 < \infty \quad \textit{and} \quad \sum_{i=1}^{\infty}\sum_{j=1}^{i-1} a_{ij}^2 < \infty. \tag{6.17}$$

If X forms a standard independent sequence with respect to the probability measure P_2, then the Radon-Nikodym density of P_2 with respect to P_1 is given by

$$\varphi(\omega) \equiv \frac{P_2(d\omega)}{P_1(d\omega)}$$

$$= \lim_{n\to\infty} \prod_{i=1}^{n} (1-\gamma_i) \exp\left[\frac{1}{2}\left\{\xi_i^2 - (1-\gamma_i)^2\left(\frac{\xi_i - \sum_{j=1}^{i-1} a_{ij}\xi_j}{1-\gamma_i}\right)^2\right\}\right]. \tag{6.18}$$

REMARK. The evaluation $\sum_i \sum_{j<i} b_{ij}^2 < \infty$ is valid for b_{ij} in the formula (6.14). This fact follows from (6.17). We can say that on the probability space $(\mathbf{\Omega}, \mathbf{B}, P_2)$, $\xi = \{\xi_i;\ i \in I^+\}$ is canonically represented in terms of $X = \{X_i;\ i \in I^+\}$ by formula (6.14). Thus, X can be viewed as the innovation process for ξ, although X is not an independent sequence.

We close this section with a short note. If X is not centered, we have (1.16′)–(1.18′) (with $E_1(X_i) = m_i$) in place of (1.16)–(1.18). That is, a Gaussian process $X = \{X_i;\ i \in I^+\}$ on a probability space $(\mathbf{\Omega}, \mathbf{B}, P_1)$ is equivalent to the standard sequence if and only if X has a canonical representation of the form

$$X_i = (1-\gamma_i)\xi_i + \sum_{j<i} a_{ij}\xi_j + m_i, \tag{6.16′}$$

where $\xi = \{\xi_i ;\ i \in I^+\}$ and where

$$\gamma_i < 1, \quad i \in I^+, \quad \Sigma\gamma_i^2 < \infty, \quad \sum_i \sum_{j<i} a_{ij}^2 < \infty, \quad \sum_i m_i^2 < \infty. \qquad (6.17')$$

In this case, the density $\varphi(\omega) = P_2(d\omega)/P_1(d\omega)$ is given by

$$\varphi(\omega) = \lim_{n\to\infty} \prod_{i=1}^{n} (1-\gamma_i) \exp\left[\frac{1}{2}\left\{\xi^2 - (1-\gamma_i)^2\left(\xi_i - (1-\gamma_i)^{-1}\sum_{j<i} a_{ij}\xi_j - m_i\right)^2\right\}\right]. \qquad (6.18')$$

§4. Canonical representations of Gaussian processes equivalent to Wiener processes

It is an interesting problem to characterize the class of continuous parameter Gaussian processes equivalent to a Brownian motion using the theory of canonical representations. The results in this section are similar in spirit to those in the previous section where we considered the case of a discrete parameter. However, since the method used to obtain a canonical representation is more complicated compared to the discrete parameter case, as in Theorem 6.2, the equivalence problem discussed here requires additional considerations.

Let $X = \{X(t);\ t \in [0, T]\}$ (T finite) be a Gaussian process on a probability space $(\Omega, \mathbf{B}, P_1)$ which is taken to be $(\mathbf{C}, \mathscr{C}, P_1)$, and assume that X is equivalent to Brownian motion. Then the Radon-Nikodym density $\varphi(\omega) > 0$ (a.e. P_1) exists, and the process X is a Brownian motion with respect to the new measure $P_2(d\omega) = \varphi(\omega)\, P_1(d\omega)$. For convenience of terminology, we say that (X, P_2) is a Brownian motion.

In this section we discuss the following steps: (i) given a Gaussian process X on $(\Omega, \mathbf{B}, P_1)$, construct a Brownian motion $B = \{B(t);\ t \in [0, T]\}$ on the same probability space as its innovation process, then (ii) form a canonical representation of X by given the canonical kernel using the Brownian motion obtained in (i). In the second step, the theory of integral kernels of Volterra type plays an important role. Our idea involves the fact that the representation theory of martingales, based on the Brownian motion (X, P_2), gives us an expression for $\varphi(\omega)^{-1} = P_1(d\omega)/P_2(d\omega)$ in terms of a stochastic integral and on the fact that the Girsanov theorem enables us to construct the canonical representation from a Brownian motion.

A brief study of stochastic integrals based on martingales is provided in Appendix §A.2. The Girsanov theorem for martingales is mentioned in §A.3.

First, we consider the density function $\varphi(\omega)^{-1} = P_1(d\omega)/P_2(d\omega)$. Define an increasing system of sub-σ-fields $\{\mathbf{B}_t(X);\ t \in [0, T]\}$ by $\mathbf{B}_t(X) = \sigma\{X(s);\ s \le t\}$. Namely $\mathbf{B}_t(X)$ is the sub-σ-field generated by the past values of X up to time t. In terms of martingale $\{X(t), \mathbf{B}_t(X);\ t \in [0, T]\}$

is a Brownian motion on the probability space $(\Omega, \mathbf{B}, P_1)$. We then define a martingale $M = \{M_t, \mathbf{B}_t(X);\ [0, T]\}$ by setting $M_t = E_2[\varphi^{-1}|\mathbf{B}_t(X)]$, where $E_2[\cdot|\cdot]$ denotes the conditional expectation on the probability space $(\Omega, \mathbf{B}, P_2)$. We will see, in the following lemma, that M has a representation in terms of a stochastic integral. Additional properties related to this topic are given in Appendix §A.3.

Lemma 6.6. (I) $P_i(M_t > 0,\ t \in [0, T]) = 1$, $i = 1, 2$.
(II) *M_t is expressed, using a stochastic integral, in the form*

$$M_t = \exp\left\{\int_0^t f(s, \omega)\, dX(s) - \frac{1}{2}\int_0^t f(s, \omega)^2\, ds\right\}, \qquad t \in [0, T], \quad (6.19)$$

where the functions $f(s, \omega)$ is
(i) *(s, ω)-measurable,*
(ii) *$\mathbf{B}_s(X)$-measurable for every fixed $s \in [0, T]$, and*
(iii) *$P_i(\int_0^T f(s, \omega)^2\, ds < \infty) = 1$, $i = 1, 2$.*

Remark. The stochastic integral (or Itô integral) of $f(s, \omega)$ is well defined if and only if the above conditions (i)–(iii) are satisfied.

Proof. It is clear that $M_T = 1/\varphi > 0$ (a.e. P_i, $i = 1, 2$). Indeed, φ is the density defined by two equivalent Gaussian measures and is strictly positive from Theorem 6.1.

We are now ready to prove (I). Introduce a stopping time τ_0:

$$\tau_0 = \begin{cases} \inf\{t;\ M_t = 0\}, & \\ T, & \text{if } M_t > 0 \text{ for every } t \in [0, T], \end{cases}$$

and a sub-σ-field by $\mathbf{B}_{\tau_0}(X) = \{A; A \in \mathbf{B},\ A \cap \{\tau_0 \le t\} \in \mathbf{B}_t(X),\ t \in [0, T]\}$. By using Theorem A.5, we obtain

$$E_2[M_t|\mathbf{B}_{\tau_0}(X)] = M_{\tau_0}, \qquad \text{a.e. } P_2.$$

In particular,

$$\int_{\{\tau_0<T\}} M_T(\omega)\, P_2(d\omega) = \int_{\{\tau_0<T\}} M_{\tau_0}(\omega)\, P_2(d\omega) = 0.$$

This means that $P_1\{\tau_0 < T\} = 0$, which implies $P_2\{\tau_0 < T\} = 0$, since the two measures P_1 and P_2 are equivalent. Thus, the first statement has been proved.

Now consider the statement (II). Since (X, P_2) is a Brownian motion, the martingale $M = \{M_t, \mathbf{B}_t(X);\ t \in [0, T]\}$ on the probability space $(\Omega, \mathbf{B}, P_2)$ with $E_2(M_t) = 1$ has the form

$$M_t = \int_0^t g(s, \omega)\, dX(s) + 1, \qquad t \in [0, T],$$

by virtue of Theorem A.9, where $g(s, \omega)$ satisfies the conditions (i), (ii), and (iii) (or belongs to the class $\mathbf{G}$ defined in §A.2.). Since $P_2(M_t > 0;$

$t \in [0, T]) = 1$ by (I), $\log M_t$ can be defined. Applying Itô's formula (Theorem A.10), we obtain the following representation of $\log M_t$:

$$\log M_t = \log\left(1 + \int_0^t g(s, \omega)\, dX(s)\right)$$
$$= \int_0^t \frac{1}{M_s} g(s, \omega)\, dX(s) - \frac{1}{2}\int_0^t \left(\frac{1}{M_s} g(s, \omega)\right)^2 ds.$$

The system of random variables $\{f(s, \omega) = g(s, \omega)/M_s,\ 0 \le s \le T\}$ also satisfies conditions (i), (ii), and (iii), so we obtain the representation (6.19). □

Since the martingales M, given in the above lemma, satisfies $E_2[M_T] = E_2[1/\varphi] = 1$, the following lemma is immediately derived from Girsanov's theorem (Theorem A.11).

LEMMA 6.7. *For the system of random variables* $\{f(s, \omega);\ 0 \le s \le T\}$ *in Lemma* 6.6, *the stochastic process* $B = \{B(t), \mathbf{B}_t(X);\ t \in [0, T]\}$, *defined by*

$$B(t) = X(t) - \int_0^t f(s, \omega)\, ds, \tag{6.20}$$

is a Brownian motion on the probability space $(\boldsymbol{\Omega}, \mathbf{B}, P_1)$.

LEMMA 6.8. *If a Gaussian process* X *defined on a probability space* $(\boldsymbol{\Omega}, \mathbf{B}, P_1)$ *is equivalent to a Brownian motion, then the random variable* $f(s, \omega)$ *in Lemma* 6.5, *satisfies*

$$E_i\left(\int_0^T f(s, \omega)^2\, ds\right) < \infty, \qquad i = 1, 2.$$

PROOF. By Theorem 6.1, the density function $M_T = 1/\varphi$ satisfies

$$E_2(M_T \log M_T) = E_1(\log M_T) = K < \infty.$$

Since the function $f(x) = x \log x$ is convex, we have the following inequality of Jensen type:

$$E_2(M_T|\mathbf{B}') \log E_2(M_T|\mathbf{B}') \le E_2(M_T \log M_T|\mathbf{B}')$$

for any sub-σ-field $\mathbf{B}' \subset \mathbf{B}$. Therefore, $f(M_t) = M_t \log M_t$ is a submartingale on the probability space $(\boldsymbol{\Omega}, \mathbf{B}, P_2)$ with respect to the increasing system of σ-fields $\{\mathbf{B}_t(X);\ t \in [0, T]\}$. Thus, for any increasing sequence of stopping times $\{T_n;\ n = 1, 2, \ldots\}$ converging to T, we obtain an inequality $E_2(M_{T_n} \log M_{T_n}) \le K$. While, from the equality

$$E_2(M_{T_n} \log M_{T_n}) = E_2(M_T \log M_{T_n}) = E_1\left(\int_0^{T_n} f\, dX(s) - \frac{1}{2}\int_0^{T_n} f^2\, ds\right)$$
$$= E_1\left(\int_0^{T_n} f\, dB(s) + \frac{1}{2}\int_0^{T_n} f^2\, ds\right)$$

we obtain an evaluation $E_1(\int_0^{T_n} f^2\, ds)/2 \le K$, provided we can choose a sequence $\{T_n;\ n = 1, 2, \dots\}$ such that $E_1(\int_0^{T_n} f(s, \omega)\, dB(s)) = 0$. Note that the existence of such a sequence is obvious. Thus, the inequality for $i = 1$ is obtained by letting n tend to ∞. As for the case $i = 2$, we note that the process $\{M_t^{-1}, \mathbf{B}_t(X);\ t \in [0, T]\}$ is a martingale on the probability space $(\Omega, \mathbf{B}, P_1)$ and therefore the same argument, as in the case $i = 1$, proves the assertion. □

We now pause to introduce some temporary notation. Let $X = \{X(t);\ t \in [0, T]\}$ be a Gaussian process on both probability spaces $(\Omega, \mathbf{B}, P_1)$ and $(\Omega, \mathbf{B}, P_2)$. The closed linear subspace of $L^2(\Omega, \mathbf{B}, P_i)$ spanned by the system $\{X(s);\ s \le t\}$ is denoted by $H_t^{(P_i)}(X)$, $i = 1, 2$.

LEMMA 6.9. *Under the same assumption as in Lemma 6.8, the random variable $f(t, \omega)$ belongs to $H_t^{(P_1)}(X)$ for each $t \in [0, T]$.*

REMARK. Since two probability spaces $(\Omega, \mathbf{B}, P_1)$ and $(\Omega, \mathbf{B}, P_2)$ are different, the square integrability conditions for a random variable with respect to two probability measures P_1 and P_2 may be different. Therefore, we must discriminate $H_t^{(P_1)}(X)$ and $H_t^{(P_2)}(X)$ at this state. We shall, however, prove the coincidence of them later (Lemma 6.10).

PROOF OF LEMMA 6.9. Set $F_s = \int_0^s f(s, \omega)\, du$. Then, we note that

$$f(s, \omega) = \lim_{h \to 0} \frac{F_{s+h} - F_s}{h} \quad \text{for almost all } s \in [0, T]. \tag{6.21}$$

This proves that

$$E_1\left\{\frac{F_{s+h} - F_s}{h}\right\}^2 = E_1\left(\int_s^{s+h} f(u, \omega)\, du/h\right)^2 \le E_1\left(\int_s^{s+h} f(s, \omega)^2\, du/h\right)$$
$$= \int_s^{s+h} \frac{\alpha(u)\, du}{h},$$

where $\alpha(u) = E_1(f(u, \omega)^2)$. Now use Lemma 6.8 to prove $\int_0^T \alpha(u)\, du < \infty$. Hence, we have

$$\lim_{h \to 0} \int_s^{s+h} \frac{\alpha(u)/du}{h} = \alpha(s) \quad \text{for almost all } s \in [0, T].$$

Therefore, the set of real numbers $\{\int_s^{s+h} \alpha(u)\, du/h;\ h \in [0, T-s]\} \cup \{\alpha(s)\}$ is bounded. This means that the system $\{(F_{s+h} - F_s)/h;\ h \in [0, T-s]\}$ is uniformly integrable on the probability space $(\Omega, \mathbf{B}, P_1)$. Use relation (6.21) to show that

$$f(s, \omega) = E_1\left(\lim_{h \to 0} \frac{F_{s+h} - F_s}{h} | \mathbf{B}_s(X)\right) = \lim_{h \to 0} \frac{E_1(F_{s+h} - F_s | \mathbf{B}_s(X))}{h}$$
$$= \lim_{h \to 0} \frac{E_1(X(s+h) - X(s) | \mathbf{B}_s(X))}{h}. \tag{6.22}$$

Since X is Gaussian,

$$E_1(X(s+h)-X(s)|\mathbf{B}_s(X)) \in H_s^{(P_1)}(X)$$

and the last limit in (6.22) converges in the mean square. Hence, we finally prove that $f(s, \omega) \in H_s^{(P_1)}(X)$. □

LEMMA 6.10. *Under the assumptions in Lemma 6.8, the two spaces $H_t^{(P_1)}(X)$ and $H_t^{(P_2)}(X)$ coincide for each $t \in [0, T]$.*

REMARK. The notation $H_t(X)$ instead of $H_t^{(P_i)}(X)$ $(i = 1, 2)$ will be used from now on.

PROOF OF LEMMA 6.10. Assume that $Z \in H_t^{(P_1)}$. Then there exists a sequence $\{Z_n;\ n = 1, 2, \dots\}$ which converges to Z in the mean square relative to the probability P_1, such that each Z_n is a linear combination of the $X(s_i)$; $s_i \in [0, t]$, $i = 1, 2, \dots, m_n$. If we choose a subsequence $\{Z_{n_k}\}$ that converges to Z almost everywhere on the probability space $(\Omega, \mathbf{B}, P_1)$, then it also converges to Z a.e. on $(\Omega, \mathbf{B}, P_2)$, because P_2 is equivalent to P_1. Since $\{X(s);\ 0 \le s \le t\}$ is Gaussian on $(\Omega, \mathbf{B}, P_2)$, so is $\{Z_{n_k};\ k = 1, 2, \dots\}$. Adding the almost everywhere convergent limit Z, the system $\{Z_{n_k};\ k = 1, 2, \dots\} \cup \{Z\}$ is still Gaussian, and therefore Z may be viewed as the limit in the mean with respect to the measure P_1. Hence, we conclude that $Z \in H_t^{(P_2)}(X)$. By interchanging P_1 and P_2, we have a similar conclusion, and hence the assertion is proved. □

LEMMA 6.11. *Under the assumption of Lemma 6.8, the random variable $f(s, \omega)$ is expressed in the form*

$$f(s, \omega) = \int_0^s k(s, u)\, dX(u) \quad \textit{for almost all } s \le T,$$

where $k(s, u)$ is a kernel of Volterra type (i.e., $k(s, u) = 0$ for $s < u$) satisfying $\int_0^t \int_0^s k(s, u)^2\, du\, ds < \infty$.

PROOF. From Lemma 6.9 and Lemma 6.10, $f(s, \omega)$ belongs to the subspace $H_s(X)$. Since the process X is a Brownian motion on $(\Omega, \mathbf{B}, P_2)$, each element Z of $H_t(X)$ can be expressed as a Wiener integral $\int_0^t k(u)\, dX(u)$, with k in $L^2([0, T])$. Therefore, we obtain the following expression

$$f(s, \omega) = \int_0^s k(s, \omega)\, dX(u)$$

for almost all s. While Lemma 6.8 guarantees

$$E_2\left\{\int_0^T \left(\int_0^s k(s, u)\, dX(u)\right)^2 ds\right\} < \infty.$$

Using the equality

$$E_2\left\{\int_0^s k(s, u)\, dX(u)^2\right\} = \int_0^s k(s, u)^2\, du,$$

and applying a measurable modification to $k(s, u)$, if necessary, we can prove that $\int_0^T \int_0^s k(s, u)^2\, du\, ds < \infty$. The kernel $k(s, u)$ is obviously of Volterra type, and the theorem is proved. □

We are now ready to prove the main theorem of this section. As will be shown below, the above lemmas, particularly Lemma 6.11, play an essential role. We keep the notation established so far in this section.

THEOREM 6.3. (I) *Let a centered Gaussian process* $X = \{X(t);\ t \in [0, T]\}$ *on a probability space* $(\Omega, \mathbf{B}, P_1)$ *be equivalent to a Brownian motion. Then on the space* $(\Omega, \mathbf{B}, P_1)$, *a Brownian motion* $B = \{B(t);\ t \in [0, T]\}$ *is formed from* X *and* X *itself has a canonical representation in terms of the* B $(H_t(X) = H_t(B),\ t \in [0, T])$ *of the following form*:

$$X(t) = B(t) - \int_0^t \int_0^s l(s, u)\, dB(u)\, ds, \qquad t \in [0, T], \tag{6.23}$$

where the kernel $l(s, u)$ *is of Volterra type and satisfies*

$$\int_0^T \int_0^s l(s, u)^2\, du\, ds < \infty.$$

Moreover, the representation (6.23) *is unique in the sense that if* X *has another canonical representation*

$$X(t) = \widetilde{B}(t) - \int_0^t \int_0^s \tilde{l}(s, u)\, d\widetilde{B}(u)\, ds,$$

then $B = \widetilde{B}$ *and* $l(s, u) = \tilde{l}(s, u)$ *for almost all* $(s, u) \in [0, T]^2$.

If the process X *is a Brownian motion on the probability space* $(\Omega, \mathbf{B}, P_2)$ *with some measure* P_2, *then there exists a Volterra kernel* $l(s, u)$ *such that the Radom-Nikodym density* $\varphi(\omega) = P_2(d\omega)/P_1(d\omega)$ *is expressed in the form*

$$\varphi(\omega) = \exp\left\{\int_0^T \int_0^s l(s, u)\, dB(u)\, dB(s) - \frac{1}{2}\int_0^T \left(\int_0^s l(s, u)\, dB(u)\right)^2 ds\right\}.$$

(II) *Conversely, if a stochastic process* X *is represented in the form* (6.23) *by A Brownian motion* B *on* $(\Omega, \mathbf{B}, P_1)$, *then it is a Gaussian process equivalent to a Brownian motion.*

PROOF. (I) A Brownian motion B on $(\Omega, \mathbf{B}, P_1)$ is constructed from X by Lemma 6.7. The Brownian motion B has a representation, by Lemma 6.11, of the following form

$$B(t) = X(t) - \int_0^t \int_0^s k(s, u)\, dX(u)\, ds.$$

Let $l(s, u)$ be the resolvent kernel of $k(s, u)$. (The existence of such a resolvent kernel follows from the theory of integral equations.) The kernel l

is also of Volterra type and is square integrable. It satisfies

$$\begin{aligned} l(s,t)+k(s,t)-\int_t^s l(s,u)k(u,t)\,du=0,\\ l(s,t)+k(s,t)-\int_t^s k(s,u)l(u,t)\,du=0 \end{aligned} \tag{6.24}$$

for almost all $s \geq t$. Using these kernels k and l, we obtain

$$\begin{aligned} B(t) &- \int_0^t\int_0^s l(s,u)\,dB(u)\,ds \\ &= X(t)-\int_0^t\int_0^s k(s,u)\,dX(u)\,ds-\int_0^t\int_0^s l(s,u)\,dX(u)\,ds \\ &\quad+\int_0^t\int_0^s l(s,u)\int_0^u k(u,v)\,dX(v)\,du\,ds \\ &= X(t)-\int_0^t\int_0^s k(s,u)\,dX(u)-\int_0^t\int_0^s l(s,u)\,dX(u)\,ds \\ &\quad+\int_0^t\int_0^s\int_v^s l(s,u)k(u,v)\,du\,dX(v)\,ds \\ &= X(t), \qquad \text{a.e. } t\in[0,T]. \end{aligned} \tag{6.25}$$

Note that the second equality comes from the fact that the order of the usual and the stochastics integral of the nonrandom integrand $k(u,v)$ is interchangeable and we have

$$\begin{aligned} \int_0^s l(s,u)\int_0^u k(u,v)\,dX(v)\,du &= \int_0^s\int_0^s l(s,u)k(u,v)\,dX(v)\,du \\ &= \int_0^s\int_v^s l(s,u)k(u,v)\,du\,dX(v). \end{aligned}$$

The third equality in (6.25) is easily obtained by (6.24). Thus, the process X has a representation in terms on B. Moreover, the representation is canonical, since $\mathbf{B}_t(X)=\mathbf{B}_t(B)$ holds for every $t \leq T$, by (6.25) and the formula established at the beginning of the proof. It is noted that this fact can also be proved by using Theorem 4.4.

The uniqueness, up to equivalence, of the Brownian motion and the kernel $l(t,s)$ comes from the uniqueness of the canonical representation. The following formula for the density function

$$\begin{aligned} \varphi(\omega) &= P_2(d\omega)/P_1(d\omega) \\ &= \exp\left\{-\int_0^T\int_0^s k(s,u)\,dX(u)\,dX(s)+\frac{1}{2}\int_0^T\left(\int_0^s k(s,u)\,dX(u)\right)^2 ds\right\}, \end{aligned}$$

comes from Lemma 6.6. Again by using (6.24) and (6.25), the above formula can be rephrased in terms of B to get the desired result.

The last step in our proof is to prove assertion (II). By Girsanov's theorem it suffices for us to show that

$$\varphi(\omega) = \exp\left\{\int_0^T \int_0^s l(s, u)\, dB(u)\, dB(s) - \frac{1}{2}\int_0^T \left(\int_0^s l(s, u)\, dB(u)\right)^2 ds\right\}$$

is a probability density, i.e., $E_1(\varphi(\omega)) = 1$. Because the process X given by the formula (6.23) is a Brownian motion on $(\Omega, \mathbf{B}, P_2)$, with $P_2(d\omega) = \varphi(\omega) P_1(d\omega)$, provided that φ is a probability density. We will therefore prove that the stochastic process $N = \{N(t);\ t \in [0, T]\}$ given below is a martingale with respect to $\{B_t(X);\ t \in [0, T]\}$ in $(\Omega, \mathbf{B}, P_1)$:

$$N_t(\omega) = \exp\left\{\int_0^t \int_0^s l(s, u)\, dB(u)\, dB(s) - \frac{1}{2}\int_0^t \left(\int_0^s l(s, u)\, dB(u)\right)^2 ds\right\};$$

or equivalently that N is a system of uniformly integrable random variables.

The uniform integrability is proved as follows. Since $N_t(\omega)$ has continuous paths with probability one, the transformed process $N_n = \{N_{t\wedge T_n},\ \mathbf{B}_t(B);\ t \in [0, T]\}$ is a martingale on $(\Omega, \mathbf{B}, P_1)$, where

$$T_n = \inf\left\{t;\ \left(\int_0^t \int_0^s l(s, u)\, dB(u)\, ds\right)^2 \geq n\right\}.$$

In particular, we have $E_1(N_{T\wedge T_n}) = 1$. Hence, if $\{N_{T\wedge T_n};\ n = 1, 2, \ldots\}$ is uniformly integrable, then we have proved

$$E_1(N_T) = E_1(\lim_{n\to\infty} N_{T\wedge T_n}) = \lim_{n\to\infty} E_1(N_{T\wedge T_n}) = 1.$$

The remainder of the proof shows that the $N_{T\wedge T_n}$ are uniformly integrable. Take the martingale N_n and define processes $X_n = \{X_n(t);\ t \in [0, T]\}$ by

$$X_n(t) = B(t) - \int_0^{t\wedge T_n} \int_0^s l(s, u)\, dB(u)\, ds.$$

From Girsanov's theorem (Theorem A.11), $X_n = \{X_n(t), \mathbf{B}_t(B);\ t \in [0, T]\}$ is a Brownian motion with respect to the probability measure $P^{(n)}$: $P^{(n)}(d\omega) = N_{T\wedge T_n}(\omega) P_1(d\omega)$ for every n. Denote by $k(s, u)$ the resolvent kernel of the Volterra kernel $l(s, u)$. Then we have

$$\int_0^s k(s, u)\, dX(u) = -\int_0^s l(s, u)\, dB(u), \quad \text{for almost all } s \in [0, T], \text{ a.e. } P_1.$$

Therefore, it follows that

$$\begin{aligned}\log N_{t\wedge T_n} &= \int_0^{t\wedge T_n}\int_0^s l(s,u)\,dB(u)\,dB(s) - \frac{1}{2}\int_0^{t\wedge T_n}\left(\int_0^s l(s,u)\,dB(u)\right)^2 ds \\ &= \int_0^{t\wedge T_n}\int_0^s l(s,u)\,dB(u)\,dX(s) + \frac{1}{2}\int_0^{t\wedge T_n}\left(\int_0^s l(s,u)\,dB(u)\right)^2 ds \\ &= -\int_0^{t\wedge T_n}\int_0^s k(s,u)\,dX(u)\,dX(s) \\ &\quad + \frac{1}{2}\int_0^{t\wedge T_n}\left(\int_0^s k(s,u)\,dX(u)\right)^2 ds \\ &= -\int_0^{t\wedge T_n}\int_0^s k(s,u)\,dX_n(u)\,dX_n(s) \\ &\quad + \frac{1}{2}\int_0^{t\wedge T_n}\left(\int_0^s k(s,u)\,dX_n(u)\right)^2 ds,\end{aligned} \tag{6.26}$$

where we have used the facts that

$$\int_0^s k(s,u)\,dX(u) = \int_0^s k(s,u)\,dX_n(u) \quad \text{for } s < T_n,$$

and

$$\int_0^t\int_0^s k(s,u)\,dx_n(u)\,dX(s) = \int_0^t\int_0^s k(s,u)\,dX_n(u)\,dX_n(s) \quad \text{for } t < T_n.$$

Since the process X_n is a Brownian motion with respect to $P^{(n)}$, (6.26) implies the following evaluations:

$$\begin{aligned}E_1(\log N_{T\wedge T_n}\cdot N_{T\wedge T_n}) &= E^{(n)}(\log N_{T\wedge T_n}) \\ &= E^{(n)}\left(\int_0^{T\wedge T_n}\int_0^s k(s,u)\,dX_n(u)\,dX_n(s)\right. \\ &\qquad\qquad \left. + \frac{1}{2}\int_0^{T_n}\left(\int_0^s k(s,u)\,dX_n(u)\right)^2 ds\right) \\ &= \frac{1}{2}E^{(n)}\left(\int_0^{T\wedge T_n}\left(\int_0^s k(s,u)\,dX(u)\right)^2 ds\right) \\ &\le \frac{1}{2}E^{(n)}\left(\int_0^T\left(\int_0^s k(s,u)\,dX_n(u)\right)^2 ds\right) = \frac{1}{2}\int_0^T\int_0^s k(s,u)^2\,du\,ds,\end{aligned}$$

where $E^{(n)}$ denotes the expectation with respect to $P^{(n)}$. Thus, we have proved the uniform integrability of $\{N_{T\wedge T_n};\ n = 1, 2, \ldots\}$, and the proof is complete. □

The remainder of this section is devoted to listing results that can be proved in the same manner as Theorem 6.3. The next theorem is concerned with a process X which is not necessarily centered.

THEOREM 6.3′. (I) *Let a Gaussian process* $X = \{X(t);\ t \in [0, T]\}$ *on a probability space* $(\Omega, \mathbf{B}, P_1) = (\mathbf{C}[0, T], \mathscr{C}, P_1)$ *be equivalent to a Brownian motion. Then a Brownian motion* B *is constructed on* $(\Omega, \mathbf{B}, P_1)$ *and* X *has a canonical representation of the form:*

$$X(t) = B(t) - \int_0^t \int_0^s l(s, u)\, dB(u)\, ds - \int_0^t a(s)\, ds, \qquad t \in [0, T], \quad (6.23')$$

where the kernel $l(s, u)$ *satisfies the same condition as in Theorem* 6.3, *and the function a belongs to* $L^2([0, T])$. *Moreover,* B, l, *and* a *are determined uniquely. If, in particular, the process* X *is a Brownian motion on the probability space* $(\Omega, \mathbf{B}, P_2)$ *with a suitable choice of* P_2, *then the Radon-Nikodym density has the form*

$$\begin{aligned}\varphi(\omega) &= \frac{P_2(d\omega)}{P_1(d\omega)} \\ &= \exp\left\{\int_0^T \left(\int_0^s l(s, u)\, dB(u) + a(s)\right) dB(s)\right. \\ &\qquad \left. - \frac{1}{2}\int_0^T \left(\int_0^s l(s, u)\, dB(u) + a(s)\right)^2 ds\right\}.\end{aligned}$$

(II) *Conversely, a stochastic process* X, *which is represented in the form* (6.23′) *by a Brownian motion* B, *is a Gaussian process equivalent to Brownian motion.*

The above two theorems deal with the equivalence problems for processes on a finite time interval $[0, T]$. However, our idea can be applied to the same problem for processes with infinite time domain, say $[0, \infty)$.

THEOREM 6.3″. (I) *Suppose that a Gaussian process* $X = \{X(t);\ t \in [0, \infty)\}$ *on a probability space* $(\Omega, \mathbf{B}, P_1) = (\mathbf{C}[0, \infty), \mathscr{C}, P_1)$ *is equivalent to a Brownian motion. Then a Brownian motion* $B = \{B(t);\ t \in [0, \infty)\}$ *is constructed on* $(\Omega, \mathbf{B}, P_1)$ *and* X *is represented in the form:*

$$X(t) = B(t) - \int_0^t \int_0^s l(s, u)\, dB(u)\, ds - \int_0^t a(s)\, ds, \qquad t \in [0, \infty), \quad (6.23'')$$

where the kernel l *and the function* a *satisfy the conditions*

$$\int_0^\infty \int_0^s l(s, u)^2\, du\, ds < \infty \quad \text{and} \quad \int_0^\infty a(s)^2\, ds < \infty. \qquad (6.27)$$

Moreover, B, l, *and* a *are determined uniquely. If, in particular, the process* X *itself is a Brownian motion on a probability space* $(\Omega, \mathbf{B}, P_2)$, *then the*

Radon-Nikodym density is expressed in the form

$$\begin{aligned}\varphi(\omega) &= \frac{P_2(d\omega)}{P_1(d\omega)} \\ &= \exp\left\{\int_0^\infty \left(\int_0^s l(s, u)\, dB(u) + a(s)\right) dB(s) \right. \\ &\qquad \left. - \frac{1}{2}\int_0^\infty \left(\int_0^\infty l(s, u)\, dB(u)\, ds + a(s)\right)^2 ds\right\}.\end{aligned}$$

(II) *Conversely, a stochastic process* X *represented in the form* (6.23″) *by a Brownian motion is a Gaussian process equivalent to a Brownian motion.*

REMARK 1. The proofs of Theorem 6.3′ and 6.3″ are obtained by a slight modification of the proof of Theorem 6.3 with an additional condition (6.27) that guarantees the regularity (or the uniform integrability) of the martingale $M = \{M_t, \mathbf{B}_t(X),\ t \in [0, \infty)\}$ on $(\Omega, \mathbf{B}, P_2)$.

REMARK 2. We have discussed, in this section, the equivalence problem on the probability space $(\Omega, \mathbf{B}, P_2)$, where Ω is taken to be $\mathbf{C}[0, T]$ or $\mathbf{C}[0, \infty)$ to be specific. However, the concept of equivalence is independent of the choice of probability spaces and such a choice is a matter of taste. If a Gaussian process X is equivalent to a Brownian motion on a probability space $(\Omega, \mathbf{B}, P_1)$, then X, roughly speaking, induces a probability distribution on $(\mathbf{C}(I), \mathscr{C})$, I being $[0, T]$ or $[0, \infty)$, because almost all sample paths of a Brownian motion are continuous. Therefore, the distribution of a process defined by (6.23) or (6.23′) or by (6.23″) may be regarded as a measure on $(\mathbf{C}(I), \mathscr{C})$, since it also has continuous paths.

§5. Canonical representation of Gaussian processes equivalent to a given Gaussian process

In this section we discuss briefly the equivalence problem for some classes of general Gaussian processes with the help of canonical representation theory. Assume that a Gaussian process $X = \{X(t);\ t \in I\}$ has a canonical representation in terms of the innovation $B = \{B(t);\ t \in I\}$ with $B(t) = (B_j(t);\ j = 1, 2, \dots, N)$, $N \le \infty$, in the form

$$X(t) = \sum_{i=1}^{N} \int_0^t F_i(t, u)\, dB_i(u), \qquad t \in I. \tag{6.28}$$

The canonical property $\mathbf{B}_t(X) = \mathbf{B}_t(B)$, $t \in I$ reduces the equivalence problem for X to that for the innovation process B which is a vector-valued process with independent increments. This comes from the fact that the martingale obtained from the densities by taking the conditional expectations have common properties related to the equivalence problem.

To be specific, we assume throughout this section that the innovation B

of X in the canonical representation (6.28) is a Brownian motion, that is;

$$B = \{B(t);\ t \in [0, T]\}, \qquad B(t) = (B_j(t);\ j = 1, 2, \dots, N),$$
$$\text{with } E(B_i(s)B_j(t)) = \delta_{ij}(s \wedge t), \qquad i, j = 1, 2, \dots, N. \quad (6.29)$$

Take a Gaussian process $Y = \{Y(t);\ t \in [0, T]\}$ on a probability space $(\Omega, \mathbf{B}, P_1)$ such that Y is equivalent to an X represented in the form (6.28). This implies that there exists a Radon-Nikodym density $\varphi(\omega)$ and that Y has the same distribution as X in the new probability space $(\Omega, \mathbf{B}, P_2)$, $P_2(d\omega) = \varphi(\omega) P_1(d\omega)$.

THEOREM 6.4. (I) *Let a centered Gaussian process Y, given on $(\Omega, \mathbf{B}, P_2)$, be equivalent to X with a canonical representation* (6.28). *Then a Gaussian process X_1 is formed on $(\Omega, \mathbf{B}, P_1)$ in such a way that the distribution of X_1 is the same as that of X and that $\mathbf{B}_t(Y) = \mathbf{B}_t(X_1)$ holds for any $t \in [0, T]$. The common innovation $B_1 = \{B_1(t);\ t \in [0, T]\}$ of Y and X_1 has the same distribution as that of B in* (6.29). *Moreover, Y has a canonical representation of the form*

$$Y(t) = \sum_{i=1}^{N} \int_0^t F_i(t, u)\, dB_i^1(u) - \int_0^t F_i(t, u) \int_0^u \sum_{j=1}^{N} l_{ij}(u, v)\, dB_j^1(v), \quad (6.30)$$

where $B_i^1(t)$ is the ith component of $B_1(t)$ and where $l_{ij}(u, v)$, $i, j = 1, 2, \dots, N$, are kernels of Volterra type: $l_{ij}(u, v) = 0$ for $u < v$ satisfying $\sum_{i,j=1}^N \int_0^T \int_0^s l_{ij}(s, u)^2\, du\, ds < \infty$.

(II) *Suppose that the process Y has the same distribution as X on $(\Omega, \mathbf{B}, P_2)$. Then the Radon-Nikodym density $\varphi(\omega) = P_2(dw)/P_1(dw)$ is expressed in the form*

$$\varphi(\omega) = \exp\left[\sum_{i=1}^{N} \int_0^T \sum_{j=1}^{N} \left(\int_0^s l_{ij}(s, u)\, dB_j^1(u)\right) dB_i^1(u) \right.$$
$$\left. -\frac{1}{2} \sum_{i=1}^{N} \int_0^T \left(\sum_{j=1}^{N} \int_0^s l_{ij}(s, u)\, dB_j^1(u)\right)^2 du\right].$$

SKETCH OF THE PROOF. Since Y has the same distribution as X with respect to the measure P_2, there exists an innovation process $B_2 = \{B_2(t);\ t \in [0, T]\}$ on the probability space $(\Omega, \mathbf{B}, P_2)$ having the same distribution as that of B in (6.29). We next determine a representation of the martingales $M = \{M_t, \mathbf{B}_t(Y);\ t \in [0, T]\}$ with $M_t = E_2(\varphi^{-1} | \mathbf{B}_t(Y))$. It is to be expressed in terms of B_2, since we have $\mathbf{B}_t(Y) = \mathbf{B}_t(B_2)$ for every t. By a method analogous to that used in Lemma 6.6, we obtain the following expression:

$$M_t = \exp\left[\sum_{i=1}^{N} \int_0^t f_i(s, \omega)\, dB_i^2(s) - \frac{1}{2} \sum_{i=1}^{N} \int_0^t f_i(s, \omega)^2\, ds\right],$$

where $B_i^2(t)$ is the ith component of $B_2(t)$. As for the random variables $f_i(s, \omega)$, we know that they are expressed in the form

$$f_i(s, \omega) = \sum_{j=1}^{N} \int_0^s k_{ij}(s, u)\, dB_j^2(u),$$

since the f_i are members of the linear space spanned by the $B_j^2(t)$. Next by the theory of integral equations of Volterra type for vector-valued functions, there exists a resolvent kernel (matrix) $(l_{ij}(s, u))$ for $(k_{ij}(s, u))$. These l_{ij} are exactly what are to be used in the representation (6.30). □

REMARK 1. As we have seen above, the equivalence problem for Y is reduced to that of the innovation B. Thus, we can develop the same argument as in §6.4, if we discuss the case where $N < \infty$. For the case of infinite multiplicity $N = \infty$, we need to extend the Itô integrals, Itô's formula, and Girsanov's theorem to obtain similar results. In fact, the arguments become more complicated in inessential parts.

REMARK 2. It is noted that the equalities

$$\mathbf{B}_t(Y) = \mathbf{B}_t(B_1) = \mathbf{B}_t(B_2) = \mathbf{B}_t(X), \qquad t \in [0, T]$$

hold, since we are always concerned with canonical representations.

REMARK 3. The expression (6.30) can be rephrased in the following manner:

$$\begin{aligned} Y(t) &= \sum_{i=1}^{N} \int_0^t \left\{ F_i(t, u) - \sum_{j=1}^{N} \int_u^t F_j(t, v) l_{ji}(v, u)\, dv \right\} dB_i^1(u) \\ &= \sum_{i=1}^{N} \int_0^t G_i(t, u)\, dB_i^1(u), \end{aligned}$$

which is the standard form for a canonical representation.

§6. Constructing innovation processes

In §4 we discussed canonical representations of Gaussian processes equivalent to a Brownian motion. The innovation problem B was squeezed in a rather straightforward manner from a given Gaussian process X, then X was represented by B. In this section, we construct an innovation from a point of view different from that of previous sections. The idea is that if a Gaussian process is given in a form "close" (in a sense) to a canonical representation, then the innovation can be constructed in a more direct manner than that in §4.

Now consider a process $X = \{X(t);\ t \in [0, T]\}$ given by

$$X(t) = B(t) + \int_0^t f(s, \omega)\, ds, \tag{6.31}$$

assuming that

(i) $\{B(t), \mathbf{F}_t; t \in [0, T]\}$ is a Brownian motion adapted to the increasing system of sub-σ-fields $\{\mathbf{F}_t; t \in [0, T]\}$, and

(ii) $f(t, \omega)$ satisfies $E\int_0^T |f(t, \omega)|\, dt < \infty$ and is $\mathbf{F}_t$-measurable for each $t \in [0, T]$. Note that X is not Gaussian in general.

Before we come to the main topic we present some examples to illustrate the idea.

EXAMPLE 1. Let $m = \{m(t, \omega); t \in [0, T]\}$ be a stochastic process independent of a Brownian motion $B = \{B(t); t \in [0, T]\}$. Set $\mathbf{B}_t(m) = \sigma\{m(s, \omega); s \leq t\}$. Assume that $f(t, \omega)$ is $\mathbf{B}_t(m)$-measurable for every $t \in [0, T]$. Define $\mathbf{F}_t$ to be $\mathbf{B}_t(B) \vee B_t(m)$, then the process $\{B(t), \mathbf{F}_t; t \in [0, T]\}$ is also a Brownian motion since the processes B and m are independent. Now define $X(t)$ by formula (6.31), and write it formally in the form

$$dX(t) = dB(t) + f(t, \omega)\, dt,$$

then we can see that it is a mathematical model of a system where a transmitted signal $f = f(t, \omega)$ coded from m is interfered with by a Gaussian white noise dB to receive dX. Such a communication system is called a system without feedback. We try to form an innovation from the observed dX, as will be discussed later.

EXAMPLE 2. Let B and m be as in Example 1, and assume that $f(t, \omega)$ is $\mathbf{F}_t$-measurable ($\mathbf{F}_t = \mathbf{B}_t(B) \vee \mathbf{B}_t(m)$). This is a system with feedback, unlike the previous example. If

$$\mathbf{B}_t(X) \subset \mathbf{F}_t, \qquad t \in [0, T], \tag{6.32}$$

is satisfied, we have an example of X of the form in (6.31). Then (6.31) can be regarded as a stochastic differential equation for an unknown process X. If the solution $X(t)$ to (6.31) satisfies the condition $X(t) \in \mathbf{F}_t$, $t \in [0, T]$, then the requirement (6.32) is automatically satisfied. As for the existence of such a solution, various kinds of sufficient conditions are known. We discuss the case of Gaussian processes in the later part of this section.

THEOREM 6.5 (A. N. Shiryaev and T. Kailath). *Let a stochastic process* $X = \{X(t); t \in [0, T]\}$ *be given by* (6.31) *with conditions* (i) *and* (ii) *for* B *and* $f(t, \omega)$. *Then, the process defined by* $B_1 = \{B_1(t), \mathbf{B}_t(X); t \in [0, T]\}$, *with*

$$B_1(t) = X_1(t) - \int_0^t g(s, \omega)\, ds, \qquad g(s, \omega) = E(f(s, \omega)|\mathbf{B}_s(X)),$$

is a Brownian motion.

PROOF. First, we note that the conditional expectation $g(s, \omega) = E(f(s, \omega)|\mathbf{B}_s(X))$ can be defined for almost all $s \in [0, T]$, and it satisfies

$$\int_0^T |g(s, \omega)|\, ds \leq \int_0^T |f(s, \omega)|\, ds < \infty.$$

For $B_1(t) = B(t) + \int_0^t (f(s, \omega) - g(s, \omega))\,ds$ note that $\mathbf{F}_s$-measurability of $f(s, \omega) - g(s, \omega)$ allows us to apply Itô's formula. In particular,

$$\begin{aligned} e^{i\lambda(B_1(t)-B_1(s))} &= 1 + i\lambda \int_s^t e^{i\lambda(B_1(u)-B_1(s))}\,dB(u) \\ &\quad + i\lambda \int_s^t e^{i\lambda(B_1(u)-B_1(s))}(f(u, \omega) - g(u, \omega))\,du \\ &\quad - \frac{\lambda^2}{2} \int_s^t e^{i\lambda(B_1(u)-B_1(s))}\,du. \end{aligned} \tag{6.33}$$

Noting that $|e^{i\lambda(B_1(u)-B_1(s))}| = 1$ and $E(\int_0^T |f - g|\,ds) < \infty$, we obtain

$$E\left(\int_s^t e^{i\lambda(B_1(u)-B_1(s))}\,dB(u)|\mathbf{B}_s(X)\right) = E(E(\cdots|\mathbf{F}_s)|\mathbf{B}_s(X)) = 0$$

and

$$\begin{aligned} &E\left(\int_s^t e^{i\lambda(B_1(u)-B_1(s))}(f - g)\,du|\mathbf{B}_s(X)\right) \\ &\quad = E\left(\int_s^t e^{i\lambda(B_1(u)-B_1(s)}E(f - g|\mathbf{B}_u(X))\,du|\mathbf{B}_s(x)\right) = 0. \end{aligned}$$

Therefore, (6.33) turns into a simple form

$$E(e^{i\lambda(B_1(t)-B_1(s))}|\mathbf{B}_s(X)) = 1 - \frac{\lambda^2}{2} \int_s^t E(e^{i\lambda(B_1(u)-B_1(s)}|\mathbf{B}_s(X))\,du,$$

from which we have the desired result

$$E(e^{i\lambda(B_1(t)-B_1(s))}|\mathbf{B}_s(X)) = \exp\left[-\frac{\lambda^2}{2}(t - s)\right] \qquad \text{a.e. } P, \ \ t, s \in [0, T].$$

Thus, B_1 is a Brownian motion. □

REMARK. In Theorem 6.5 it is obvious that $\mathbf{B}_t(X) \supset \mathbf{B}_t(B_1)$, however the converse inclusion "$\subset$" is not always true. In the case that $\mathbf{B}_t(X) = \mathbf{B}_t(B_1)$ for every $t \in [0, T]$, the Brownian motion B_1 is the innovation of X.

Next we apply the Theorem to a Gaussian communication systems. Suppose that $k(s, u)$ is a Volterra kernel satisfying the condition

$$\int_0^T \int_0^s k(s, u)^2\,du\,ds < \infty,$$

and assume that

$$m = \{m(t, \omega);\ t \in [0, T]\}$$

is a Gaussian process independent of a Brownian motion B, satisfying $E(\int_0^T m(s, \omega)^2\,ds) < \infty$. Define $X = \{X(t);\ t \in [0, T]\}$ by

$$X(t) = B(t) + \int_0^t \left(\int_0^s k(s, u)\,dX(u) + m(s, \omega)\right) ds. \tag{6.34}$$

For such a Gaussian process X, m is the message, dB is the noise in the system and $\int_0^s k(s, u)\, dX(u)$ is expressing a linear feedback that is $\mathbf{B}_s(X)$-measurable. If the stochastic differential equation (6.34) for X has a $\mathbf{B}_t(B) \vee \mathbf{B}_t(m)$-measurable solution, the process X has the form in (6.31). Therefore, the process $B_1 = \{B_1(t), \mathbf{B}_t(X);\ t \in [0, T]\}$ defined by

$$\begin{aligned} B_1(t) &= X(t) - \int_0^t E\left(\int_0^s k(s, u)\, dX(u) + m(s, \omega) | \mathbf{B}_s(X)\right) ds \\ &= B(t) + \int_0^t \{m(s, \omega) - E(m(s, \omega)|\mathbf{B}_s(X))\}\, ds, \qquad t \in [0, T], \end{aligned} \tag{6.35}$$

is a Brownian motion by Theorem 6.5. By using this fact, we can prove the existence and uniqueness of the solution of (6.34). Set $Y(t) = B(t) + \int_0^t m(s, \omega)\, ds$, then $Y = \{Y(t);\ t \in [0, T]\}$ is equivalent to to a Brownian motion. In fact, the expectation of $\alpha_T(\omega)$ is equal to equal to 1, because of the independence of B and m, where

$$\alpha_t(\omega) = \exp\left\{-\int_0^t m(s, \omega)\, dB(s) - \frac{1}{2}\int_0^t m(s, \omega)^2\, ds\right\}, \qquad t \in [0, T].$$

Applying Girsanov's theorem (Theorem A.11), the process $Y = \{Y(t), \mathbf{B}_t(B) \vee \mathbf{B}_t(m);\ t \in [0, t]\}$ is a Brownian motion on the probability space $(\mathbf{\Omega}, \mathbf{B}, P_1)$, with $P_1(d\omega) = \alpha_T(\omega) P(d\omega)$. Since (6.34) can be written as an integral equation in X, i.e.,

$$X(t) = Y(t) + \int_0^t \int_0^s k(s, u)\, dX(u)\, ds, \qquad t \in [0, T],$$

it has a unique solution in the form

$$X(t) = Y(t) - \int_0^t \int_0^s l(s, u)\, dY(u)\, ds, \qquad t \in [0, T], \tag{6.36}$$

by using the resolvent kernel $l(s, u)$ of $k(s, u)$ as in §6.4. From Theorem 6.3, we see that X is also equivalent to a Brownian motion and (6.36) is the canonical representation of X by the Brownian motion Y on $(\mathbf{\Omega}, \mathbf{B}, P_1)$. Thus, we have $\mathbf{B}_t(X) = \mathbf{B}_t(Y) \subset \mathbf{B}_t(B) \vee \mathbf{B}_t(m)$, $t \in [0, T]$. Therefore, we see that equation (6.34) has a unique solution. Since the triple (B, m, X) is a vector-valued Gaussian process, the conditional expectation $E(m(s, \omega)|\mathbf{B}_s(x))$ is a member of the linear span $H_t(X)$ of $\{X(s);\ s \in [0, t]\}$. By Lemma 6.11, there exists a Volterra kernel $h(s, u)$ satisfying $\int_0^T \int_0^s h(s, u)^2\, du\, ds < \infty$, and we have

$$E(m(s, \omega)|\mathbf{B}_s(X)) = \int_0^s h(s, u)\, dX(u), \qquad \text{a.e. } P,\ s \in [0, T].$$

Therefore, we have from (6.35) that

$$B_1(t) = X(t) - \int_0^t \int_0^s (k(s, u) + h(s, u))\, dX(u)\, du.$$

Again by a discussion similar to that in §4, we can find the resolvent kernel $n(s, u)$ of $j(s, u) = k(s, u) + h(s, u)$. Using this kernel $n(s, u)$, X has a representation of the form

$$X(t) = B_1(t) - \int_0^t \int_0^s n(s, u)\, dB_1(u)\, ds.$$

The above arguments yield the following corollary.

COROLLARY. *If a Gaussian process X is given by* (6.34), *then the Brownian motion B_1 in* (6.35) *is the innovation of X.*

REMARK. In general, the condition $\int_0^T m(s, \omega)^2\, ds < \infty$ is equivalent to $E(\int_0^T m(s, \omega)^2\, ds) = \int_0^T E(m(s, \omega)^2)\, ds < \infty$. This relation has been used in this section.

APPENDIX

Stochastic Integrals and Martingales

In this Appendix we introduce some basic notions about stochastic integrals and martingales on a Wiener space. Results presented here have been used in this book, in particular in §6.4.

§1. Multiple Wiener integrals

The Wiener space $(C, \mathbf{C}, \mu_W)$ is taken to be the basic probability space $(\Omega, \mathbf{B}, P)$, in what follows. The set C is the space of continuous functions on R^1, and $\mathbf{C}$ is the σ-field generated by cylinder subsets of C. The measure μ_W is the probability distribution of a standard Brownian motion $B = \{B(t)\,;\ t \in [0, \infty)\}$ introduced in §2.5. The Wiener integral is defined in §2.5 and is denoted by $\int_0^t f(s)\,dB(s)$, for $f \in L^2([0, \infty))$. As a generalization of this integral, we are going to define, in this section, a multiple Wiener integral, which may be thought of as an analogue of the multiple integral on a higher dimensional Euclidean space defined as a generalization of the ordinary integral on R^1. The direct sum decomposition of $L^2(\Omega) = L^2(\Omega, \mathbf{B}, P)$ or the Fock space will be given by using multiple Wiener integrals.

1°. Definition of multiple Wiener integrals. Let $n > 0$ be fixed. Suppose that $k = k(t_1, t_2, \dots, t_n)$ is in the space

$$L^2(R_+^n) = \left\{k\,;\ \int_0^\infty \int_0^{t_1} \cdots \int_0^{t_{n-1}} k(t_1, t_2, \dots, t_n)^2\, dt_1\, dt_2 \cdots dt_n < \infty\right\}.$$

We can assume that $k(t_1, t_2, \dots, t_n) = 0$, if $t_i < t_j$ for some $i < j$, for convenience. Such a kernel k is called an (n-ple) *Volterra kernel*. The Wiener integral of a Volterra kernel k is defined in the following steps.

The first step. The case where k is a step function: there exists instants $0 = \tau_0 < \tau_1 < \tau_2 < \cdots < \tau_r$ such that

$$\begin{aligned} k(t_1, t_2, \dots, t_n) &= a_{i_1, i_2, \dots, i_n}, \qquad \tau_{i_1} \le t_1 < \tau_{i_1+1},\ \tau_{i_2} \le t_2 < \tau_{i_2+1}, \\ &\qquad \dots, \tau_{i_n} < t_n < \tau_{i_n+1}, \quad r > i_1 > i_2 > \cdots > i_n, \\ &= 0, \quad \text{otherwise}. \end{aligned}$$

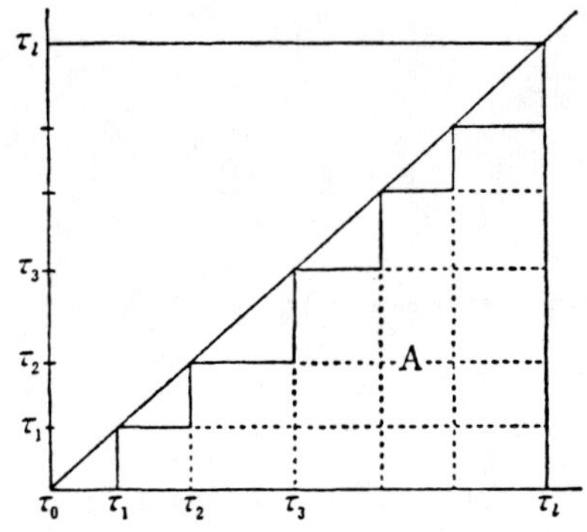

FIGURE A.1

For this k, the *n-ple Wiener integral* $I_n(k)$ is defined by

$$\begin{aligned} I_n(k) &\equiv \iint \cdots \int k(t_1, t_2, \dots, t_n)\, dB(t_1)\, dB(t_2) \cdots dB(t_n) \\ &= \sum_{r>i_1>i_2>\cdots>i_n} \sum \cdots \sum a_{i_1, i_2, \dots, i_n} (B(\tau_{i_1+1}) - B(\tau_{i_1})) \\ &\qquad \cdot (B(\tau_{i_2+1}) - B(\tau_{i_2})) \cdots (B(\tau_{i_n+1}) - B(\tau_{i_n})). \quad \text{(A.1)} \end{aligned}$$

It is easy to obtain the mean and the convariance of multiple Wiener integrals in this case. Take two random variables $I_n(k)$ and $I_m(l)$ expressed in the form (A.1). See Figure A.1. There we may assume without loss of generality that both step functions k and l have a common division of $R^1: 0 \le \tau_1 < \tau_2 < \cdots < \tau_q$. Since a Brownian motion has independent increments and satisfies the equalities $E((B(\tau_{i+1}) - B(\tau_i))^2) = \tau_{i+1} - \tau_i$, $i = 0, 1, 2, \dots, q-1$, we have

$$\begin{aligned} E(I_n(k)) &= 0, \qquad n \ge 1 \\ E(I_n(k) I_m(l)) &= \int_0^\infty \int_0^{t_1} \cdots \int_0^{t_{m-1}} k(t_1, t_2, \dots, t_m) \\ &\quad \cdot l(t_1, t_2, \dots, t_m)\, dt_m \cdots dt_2\, dt_1 = (k, l)_n \quad \text{if } n = m, \\ &= 0 \quad \text{if } n \ne m. \end{aligned} \tag{A.2}$$

Here $(\cdot, \cdot)_n$ denotes the inner product in $L^2(R_+^n)$. It can be proved that the variance of $I_n(k)$ is finite, or $I_n(k) \in L^2(\Omega)$.

The second step. For a given general Volterra kernel $k = k(t_1, t_2, \dots, t_n) \in L^2(R_+^n)$, we can find a sequence $\{k_i(t_1, t_2, \dots, t_n);\ i = 1, 2, \dots\}$ of step functions (kernels) that approximate k in $L^2(R_+^n)$: $\lim_{i\to\infty} \int \cdots \int |k(t_1, \dots, t_n) - k_i(t_1, \dots, t_n)|^2\, dt_n \cdots dt_1 = \lim_{i\to\infty} \|k - k_i\|_n^2 = 0$. Since the sequence $\{k_i\}$ satisfies the Cauchy condition, we obtain

$$E(|I_n(k_i) - I_n(k_j)|^2) = \|k_i - k_j\|_n^2 \to 0 \quad \text{as } i, j \to \infty.$$

Thus, there exists a limit in $L^2(\Omega)$, which is denoted by $I_n(k)$:

$$\operatorname*{l.i.m.}_{i\to\infty} I_n(k_i) = I_n(k). \tag{A.3}$$

The random variable $I_n(k)$ defined above is called the n-ple *Wiener integral* of k and is denoted by

$$I_n(k) = \int\int\cdots\int k(t_1, t_2, \ldots, t_n)\, dB(t_n)\cdots dB(t_2)\, dB(t_1).$$

Note that $I_n(k)$ is well defined, namely the limit (A.3) is independent of the choice of a sequence $\{k_i\}$.

Next we give some properties of multiple Wiener integrals.

Theorem A.1. (i) (linearity) *Let k_1 and k_2 be two Volterra kernels in $L^2(R^n_+)$. Then,*

$$I_n(\alpha k_1 + \beta k_2) = \alpha I_n(k_1) + \beta I_n(k_2), \qquad \alpha, \beta \text{ constants.}$$

(ii) *Equality* (A.2) *is valid for any two Volterra kernels $k \in L^2(R^n_+)$ and $l \in L^2(R^m_+)$.*

Proof. Both (i) and (ii) are trivial for step functions and these properties are inherited by their limits because of (A.2) and (A.3). □

For any n, the set of all n-ple Wiener integrals forms a subspace of $L^2(\Omega)$, which is denoted by $\mathscr{H}_n$:

$$\mathscr{H}_n = \{I_n(k);\ k \in L^2(R^n_+),\ k \text{ is a Volterra kernel}\}.$$

We tacitly set $\mathscr{H}_0 = C$ (the complex number field). In the case $n \neq m$, $\mathscr{H}_n$ and $\mathscr{H}_m$ are mutually orthogonal. In particular, $\mathscr{H}_1$ is the subspace of Wiener integrals defined previously and it is in agreement with the linear span $H(B)$ obtained from the Brownian motion B.

A family of subspaces $\mathscr{H}_n(t)$, $t \in [0, \infty)$, are now introduced:

$$\mathscr{H}_n(t) = \{I_n(k) \in \mathscr{H}_n;\ k(t_1, t_2, \ldots, t_n) = 0,\ t_1 < t\}, \qquad n = 1, 2, \ldots.$$

Theorem A.2. *For any $n = 0, 1, 2, \ldots$, and $t \in [0, \infty)$, we have*

$$E(I_n(k)|\mathbf{B}_t(B)) = I_n(k^t), \tag{A.4}$$

where

$$k^t(t_1, \ldots, t_n) = \begin{cases} k(t_1, \ldots, t_n), & t_1 \le t; \\ 0, & t_1 > t. \end{cases}$$

Namely, $I_n(k^t)$ is the orthogonal projection of $I_n(k)$ to $\mathscr{H}_n(t)$.

PROOF. For a step function k, the result is obvious by noting the property of conditional expectation is Proposition 1.3. In the general case, (A.4) can be proved by the fact that the convergence of $\{k_i\}$ in $L^2(R_+^n)$ implies that of $\{k_i^t\}$ in $L^2(R_+^n)$. □

NOTATION. We denote $I_n(k^t)$ in (A.4) by

$$\int_0^t \int_0^{t_1} \cdots \int_0^{t_n} k(t_1, t_2, \ldots, t_n)\, dB(t_n) \cdots dB(t_2)\, dB(t_1).$$

2°. Completeness of multiple Wiener integrals. The space $L^2(\Omega) = L^2(\Omega, \mathbf{B}, P)$ of all random variables with finite variance on $(\Omega, \mathbf{B}, P) = (C, \mathbf{C}, \mu_W)$ admits a direct sum decomposition into the sum of the spaces $\mathscr{H}_n$, $n \geq 0$, of multiple Wiener integrals, i.e., any $f \in L^2(\Omega)$ can be expanded into a series of multiple Wiener integrals. The proof of this fact uses the Hermite polynomials defined below:

DEFINITION A.1. A polynomial in (t, x), $t \geq 0$, $x \in R$, given by

$$H_n(t, x) = \frac{(-t)^n}{n!} \exp\left[\frac{x^2}{2t}\right] \frac{d^n}{dx^n} \exp\left[\frac{-x^2}{2t}\right], \qquad n = 0, 1, 2, \ldots,$$

is called a *Hermite polynomial of degree* n (with parameter t).

The following Proposition lists some useful properties of Hermite polynomials.

PROPOSITION A.1. (i) *The generating function of the* H_n *is*

$$\sum_{n=0}^{\infty} \lambda^n H_n(t, x) = \exp\left[\lambda x - \frac{\lambda^2 t}{2}\right].$$

(ii) *For* $n \geq 2$,

$$H_n(t, x) - \frac{x}{n} H_{n-1}(t, x) + \frac{t}{n} H_{n-2}(t, x) = 0 \text{ (\textit{the recursion formula})}.$$

(iii) *The system* $\{H_n(t, x);\ n = 0, 1, 2, \ldots\}$ *is complete and orthogonal in the Hilbert space* $L^2((-\infty, \infty), N_t) \equiv \{f;\ \int f(x)^2 N_t(dx) < \infty\}$, *where* $N_t(dx) = (\exp[-x^2/2t]/\sqrt{2\pi t}\, dx$. *The normalized system is of the form*

$$\left\{\frac{\sqrt{n!}}{\sqrt{t}} H_n(t, x);\ n \geq 0\right\}.$$

(iii′) *The system*

$$\{H_{n_1}(t_1, x_1) H_{n_2}(t_2, x_2) \cdots H_{n_d}(t_d, x_d);\ n_1, n_2, \ldots, n_d = 0, 1, 2, \ldots\}$$

is a complete orthogonal system in the Hilbert space

$$L^2(R^d, N_{(t_1, t_2, \dots, t_d)}(dx_1\, dx_2 \cdots dx_d)),$$

where $N_{(t_1, t_2, \dots, t_d)}$ *is a Gaussian measure on* R^d *with mean vector zero and covariance matrix* $(v_{ij}) = (t_i \delta_{ij})$.

The proof comes from elementary properties of the Hermite polynomials and so is omitted.

Incidentally, the usual Hermite polynomial denoted by $H_n(x)$ is $n!2^{n/2} H_n(\sqrt{2}x; 1)$.

LEMMA A.1. *For any* $n \geq 0$ *and for any* s, t $(s \leq t)$, *we have*

$$\begin{aligned} H_n(t-s, B(t)-B(s)) &= \int_s^t \int_s^{t_1} \cdots \int_s^{t_{n-1}} dB(t_n) \cdots dB(t_2)\, dB(t_1) \\ &= \iint \cdots \int k(t_1, t_2, \dots, t_n)\, dB(t_n) \cdots dB(t_2)\, dB(t_1), \end{aligned} \tag{A.5}$$

where $k(t_1, t_2, \dots, t_n) = 1$, *for* $t \geq t_1 \geq t_2 \geq \cdots \geq t_n \geq s$; *and is* 0, *otherwise.*

OUTLINE OF THE PROOF. $H_0(t-s, B(t)-B(s)) = 1$ and $H_1(t-s, B(t)-B(s)) = B(t)-B(s) = \int_s^t dB(u)$ are obvious. We now use mathematical induction. Assume that

$$H_k(t-s, B(t)-B(s)) = \int_s^t \int_s^{t_1} \cdots \int_s^{t_{k-1}} dB(t_k) \cdots dB(t_2)\, dB(t_1),$$

for $k = 1, 2, \dots, n-1$. We then form a sequence that approximates $\int_s^t \int_s^{t_1} \cdots \int_s^{t_{n-1}} dB(t_n) \cdots dB(t_2)\, dB(t_1)$. For a division Δ of $[s, t]$: $\tau = s < \tau_1 < \tau_2 < \cdots < \tau_q = t$,

$$\begin{aligned} &\sum_{q \geq i_1 > i_2 > \cdots > i_n \geq 1} \sum \cdots \sum (B(\tau_{i_1}) - B(\tau_{i_1 - 1}))(B(\tau_{i_2}) - B(\tau_{i_2 - 1})) \cdots (B(\tau_{i_n}) - B(\tau_{i_n - 1})) \\ &= \frac{1}{n} \left(\sum_{k=1}^{q} \left((B(\tau_k) - B(\tau_{k-1})) \sum_{\substack{q \geq i_1 > i_2 > \cdots > i_{n-1} \geq 1 \\ i_1, i_2, \dots, i_{n-1} \neq k}} \cdots \sum (B(\tau_{i_1}) - B(\tau_{i_1 - 1})) \right.\right. \\ &\qquad \left.\left. \cdots (B(\tau_{i_{n-1}}) - B(\tau_{i_{n-1}-1})) \right) \right) \end{aligned}$$

$$= \frac{1}{n}\left(\sum_{k=1}^{q}(B(\tau_k) - B(\tau_{k-1})) \sum_{q\geq i_1>i_2>\cdots>i_{n-1}\geq 1}\cdots\sum (B(\tau_{i_1}) - B(\tau_{i_1-1})) \cdots (B(\tau_{i_{n-1}}) - B(\tau_{i_{n-1}-1}))\right.$$

$$- \sum_{k=1}^{q}(B(\tau_k) - B(\tau_{k-1}))^2 \left(\sum_{q\geq i_1>\cdots>i_{n-2}\geq 1}\cdots\sum (B(\tau_{i_1}) - B(\tau_{i_1-1})) \cdots (B(\tau_{i_{n-2}}) - B(\tau_{i_{n-2}-1}))\right)$$

$$\left. + \left(\sum_{k=1}^{q}(B(\tau_k) - B(\tau_{k-1}))^3 \sum_{\substack{q\geq i_1>\cdots>i_{n-3}\geq 1\\ i_1,\dots,i_{n-3}\neq k}}\cdots\sum (B(\tau_{i_1}) - B(\tau_{i_1-1})) \cdots (B(\tau_{i_{n-3}}) - B(\tau_{i_{n-3}-1}))\right)\right). \tag{A.6}$$

The second and the third term of the right-hand side of (A.6) are the rectifications of $(B(\tau_k)-B(\tau_{k-1}))^\alpha$, $\alpha = 2, 3$, in the first term. As the division Δ becomes finer, the left-hand side of (A.6) converges to $\int_s^t \int_s^{t_1} \cdots \int_s^{t_{n-1}} dB(t_n) \cdots dB(t_2)\, dB(t_1)$ in $L^2(\Omega)$. While,

$$\sum_{k=1}^{q}(B(\tau_k) - B(\tau_{k-1})) = B(t) - B(s),$$

$$\sum_{k=1}^{q}(B(\tau_k) - B(\tau_{k-1}))^2 \to t - s, \quad \text{and} \quad \sum_{k=1}^{q}(B(\tau_k) - B(\tau_{k-1}))^3 \to 0,$$

as are well known. From these facts, the right-hand side of (A.6) converges to

$$\frac{1}{n}(B(t) - B(s)) \int_s^t \int_s^{t_1} \cdots \int_s^{t_{n-2}} dB(t_{n-1}) \cdots dB(t_2)\, dB(t_1)$$
$$- \frac{1}{n}(t - s) \int_s^t \int_s^{t_1} \cdots \int_s^{t_{n-3}} dB(t_{n-2}) \cdots dB(t_2)\, dB(t_1).$$

Using the assumption and Proposition A.1(ii), , we can prove the equalities in (A.5) for $k = n$. □

The expression established above is generalized similarly.

COROLLARY. *For any* $n \geq 0$, *and for* $f \in L^2([0, \infty))$,

$$
\begin{aligned}
H_n\left(\int_0^\infty f(s)^2\,ds, \int_0^\infty f(s)\,dB(s)\right) &= \int_0^\infty \int_0^{t_1} \cdots \int_0^{t_{n-1}} f(t_n)\,dB(t_n)\cdots f(t_2)\,dB(t_2) f(t_1)\,dB(t_1) \\
&= \iint \cdots \int k(t_1, t_2, \dots, t_n)\,dB(t_n)\cdots dB(t_2)\,dB(t_1), \qquad \text{(A.5}')
\end{aligned}
$$

where

$$
k(t_1, t_2, \dots, t_n) = \begin{cases} f(t_1)f(t_2)\dots f(t_n) & \text{for } t_1 \geq t_2 \geq \cdots \geq t_n, \\ 0 & \text{otherwise.} \end{cases}
$$

LEMMA A.2. *Let* $\Lambda = \{\tau_i;\ i = 1, 2, \dots\}$ *be a countable dense subset of the time parameter set* $I = [0, \infty)$ *and let* $\mathbf{B}_\Lambda$ *be a sub-*σ*-field generated by* $\{\omega(\tau_i) - \omega(\tau_j);\ i, j = 1, 2, \dots\}$, *where* $\omega \in C$. *Then* $\mathbf{B} = \mathbf{B}_\Lambda$.

PROOF. Recall that the probability space $(\Omega, \mathbf{B}, P)$ is taken to be $(C, \mathbf{C}, \mu_W)$. Since the σ-algebra $\mathbf{B} = \mathbf{C}$ is generated by cylinder sets, it suffices to prove that $\mathbf{B}\{t\} = \sigma\{\omega(t)\} \equiv \sigma$-algebra generated by $\{\omega(t)\}$ is in $\mathbf{B}_\Lambda$ for t outside of Λ. Since $\mathbf{B}\{t\}$ is generated by the family of sets $\{\omega\colon \omega(t) \in O,\ O \text{ is an open set}\}$, it is sufficient to prove $\{\omega(t) \in O\} \in \mathbf{B}_\Lambda$. Let $\{\tau_{i_k}\}$ be a sequence converging to t. Then, $\{\omega(t) \in O\} = \bigcup_{k=1}^\infty \bigcap_{j>k} \{\omega(\tau_{i_j}) \in O\} \in \mathbf{B}_\Lambda$ is shown. $\square$

THEOREM A.3. *A random variable* $f = f(\omega) \in L^2(\Omega) = L^2(\Omega, \mathbf{B}, P)$ *has an expansion of the form*

$$
\begin{aligned}
f(\omega) &= \sum_{i=0}^\infty \int \cdots \iint k_i(t_1, t_2, \dots, t_i)\,dB(t_i)\cdots dB(t_2)\,dB(t_1) \\
&= \sum_{i=0}^\infty I_i(k_i), \qquad \sum_{i=0}^\infty \|k_i\|_i^2 < \infty, \qquad \text{(A.7)}
\end{aligned}
$$

in terms of multiple Wiener integrals. Or equivalently, $L^2(\Omega)$ *is decomposed into*

$$
L^2(\Omega) = \sum_{i=0}^\infty \bigoplus \mathscr{H}_i \ (\textit{direct sum}), \quad \textit{where } \mathscr{H}_0 = R.
$$

PROOF. For the finite subsets $\Lambda_l = \{\tau_1, \tau_2, \dots, \tau_l\}$, $l = 1, 2, \dots$, of Λ in Lemma A.2 define a sub-σ-field $\mathbf{B}_l$ by $\sigma\{\omega(\tau_i)-\omega(\tau_j);\ i, j=1, 2, \dots, l\}$. From the results in Lemma A.2, $\mathbf{B} = \mathbf{B}_\Lambda = \bigvee_l \mathbf{B}_l$. Therefore, $f_l = E(f|\mathbf{B}_l)$, with $f \in L^2(\Omega)$, converges in the mean to f. Indeed, $\{f_l;\ l = 1, 2, \dots\}$

is a regular martingale and $\{f_l^2;\ l = 1, 2, \dots\}$ is uniformly integrable. Having arranged the members of Λ_l in order $s_1 < s_2 < \cdots < s_l$, we can say that $\mathbf{B}_l$ is generated by $\{\omega(s_{k+1}) - \omega(s_k);\ 1 \le k \le l\}$. The definition of the Brownian motion B on the measure space $(\Omega, \mathbf{B}, P) = (C, \mathbf{C}, P)$ allows us to take $B(t, \omega) = \omega(t)$, $t \in I$. Each f_l is a square integrable function of the independent random variables $\{(B(s_{k+1}) - B(s_k);\ k = 1, \dots, l-1\}$. Therefore, Proposition A.1 (iii) enables us to derive the expansion

$$f_l = \sum_{i_1, \dots, i_{l-1}} \cdots \sum a_{i_1 \cdots i_{l-1}} H_{i_1}(s_2 - s_1, B(s_2) - B(s_1)) \\ \cdots H_{i_{l-1}}(s_l - s_{l-1}, B(s_l) - B(s_{l-1})).$$

This implies that $f_l \in \sum_{i=0}^{\infty} \oplus \mathscr{H}_i$. Since $\sum_{i=0}^{\infty} \oplus \mathscr{H}_i$ is a closed subspace of $L^2(\Omega)$, the limit f also belongs to this subspace. □

COROLLARY. *The expansion* (A.7) *is unique: Volterra kernels* $k_i(t_1, \dots, t_i)$, $i = 1, 2, \dots$, *are uniquely determined by* $f \in L^2(\Omega)$.

DEFINITION A.2. The formula (A.7) is called the *Wiener expansion* of $f \in L^2(\Omega)$.

§2. Martingales and stochastic (Itô) integrals

1°. Martingales with continuous parameter. Based on the results for discrete parameter martingales §1.5, we will discuss martingales with continuous parameter, which we take to be an interval $I = [0, T]$, $T < \infty$, or $I = [0, \infty)$.

DEFINITION A.3. A stochastic process $M = \{M_t(\omega);\ t \in I\}$ on a probability space is said to be a *martingale with respect to an increasing system of sub-σ-fields* $\{\mathbf{F}_t;\ t \in I\}$, if the conditions

(M.1) $E(|M_t|) < \infty$, $t \in I$,

(M.2) $E(M_t | \mathbf{F}_s) = M_s$, a.e. P for $s \le t$.

REMARK. Sometimes the notation $M = \{M_t, \mathbf{F}_t;\ t \in I\}$ is used for a martingale to specify the σ-fields.

DEFINITION A.4. A stochastic process M is said to be a *submartingale* (resp. *supermartingale*) with respect to $\{\mathbf{F}_t;\ t \in I\}$, if it satisfies the conditions (M.1) and

$$E(M_t | \mathbf{F}_s) \ge (\text{resp.}, \le) M_s \quad \text{a.e. } P,\ s \le t. \tag{$M.2'$}$$

As in the discrete parameter case, a martingale $M = \{M_t, F_t;\ t \in I\}$ is called *regular*, if there exists a random variable $F(\omega) \in L^1(\Omega) \equiv \{F;\ E(|F|) < \infty\}$ satisfying $M_t = E(F | \mathbf{B}_t)$. For such an M_t we see that the sample path is continuous.

DEFINITION A.5. If sample paths $M_t(\omega)$ of a (sub or super) martingale are continuous (resp., right continuous) a.e., then M is said to be *continuous* (resp. *right continuous*).

REMARK. In 2° of this section, we will see that all martingales with respect to $\{\mathbf{B}_t(B);\ t \in I\}$ (B; a Brownian motion) are continuous.

EXAMPLE 1. Let $f(x)$ be a convex function. Then, $\{f(M_t),\ \mathbf{F}_t;\ t \in I\}$ is a submartingale, if $M = \{M_t, \mathbf{F}_t;\ t \in I\}$ is martingale. This fact is proved by using Jensen's inequality as in the discrete parameter case.

THEOREM A.4 (Doob's inequality). *Let* $M = \{M_t, \mathbf{F}_t;\ t \in [0, T]\}$ *be a right continuous martingale* (*or submartingale*). *Then the inequalities*

$$P\left(\sup_{t\le T} M_t \ge \lambda\right) \le \frac{1}{\lambda}\int_{\{\sup_{t\le T} M_t \ge \lambda\}} M_T\, P(d\omega) \le \frac{1}{\lambda} E(M_T^+), \qquad M_T^+ \equiv M_T \vee 0, \tag{A.8}$$

and

$$P\left(\inf_{t\le T} M_t \le -\lambda\right) \le \frac{1}{\lambda}\left\{-E(M_0) + \int_{\{\inf_{t\le T} M_t \ge -\lambda\}} M_T\, P(d\omega)\right\} \tag{A.9}$$

hold for all $\lambda > 0$.

PROOF. Denote the set of all rational numbers in $[0, t]$ by $\mathbf{Q} = \{r_i;\ i = 1, 2, \dots\}$. Consider a restricted parameter process $M^0 = \{M_r^0, \mathbf{B}_r;\ r \in \mathbf{Q}\}$. The right continuity of M yields equalities

$$P\left(\sup_{t\le T} M_t \ge \lambda\right) = P\left(\sup_{r\in\mathbf{Q}} M_r \ge \lambda\right)$$

and

$$P\left(\inf_{t\le T} M_t \le -\lambda\right) = P\left(\inf_{r\in\mathbf{Q}} M_r \le -\lambda\right).$$

Take a subset $\mathbf{Q}_k = \{r_i;\ i = 1, 2, \dots\}$ of $\mathbf{Q}$ and arrange in increasing order $r_{i_1} < t_{i_2} < \cdots < r_{i_k}$. Then $M^{(k)} = \{M_{r_i}, \mathbf{B}_{r_i};\ i = 1, 2, \dots, k\}$ is a (sub)martingale with discrete parameter, and Doob's inequality for $M^{(k)}$ holds (Theorem 1.10). Thus, the relation

$$\left\{\omega;\ \sup_{r\in\mathbf{Q}} M_r(\omega) \ge \lambda\right\} = \bigcap_{n=1}^{\infty}\bigcup_{k=1}^{\infty}\left\{\omega;\ \max_{i=1,\dots,k} M_{r_i}(\omega) \ge \lambda - \frac{1}{n}\right\}$$

proves (A.8). (A.9) is proved using the same argument. □

The stopping time is defined in a manner similar to the discrete parameter case.

DEFINITION A.6. Let $\tau = \tau(\omega)$ be an $I \cup \{\infty\}$-valued random variable. Then, τ is called a *stopping time* with respect to an increasing system of

σ-fields $\{\mathbf{F}_t;\ t \in I\}$ if the event $\{\omega;\ \tau(\omega) \leq t\}$ belongs to $\mathbf{F}_t$ for each $t \in I$. For such a τ, the sub-σ-field $\{A;\ A \cap \{\omega;\ \tau(\omega) \leq t\} \in \mathbf{F}_t,\ t \in I\}$ is denoted by $\mathbf{F}_\tau$.

For two stopping times σ and τ, if $\sigma \leq \tau$, a.e. P, then $\mathbf{F}_\sigma \subset \mathbf{F}_\tau$.

EXAMPLE 2. (1) Each $t \in I$ is a (deterministic) stopping time.

(2) If $M = \{M_t, \mathbf{F}_t;\ t \in I\}$ is a continuous (sub)martingale, then the hitting time $\tau = \inf\{t;\ M_t = a\}$ to a point $a \in \mathbf{R}$ is a stopping time, where we set $\inf \phi = \sup I$.

(3) Let $f(s, \omega)$ be jointly measurable in $(s, \omega) \in I \times \Omega$ and be integrable in s, almost surely. Then $\tau = \inf\{t;\ \int_0^t f(s, \omega)\,ds \geq a\}$ is a stopping time.

LEMMA A.3. *If $M = \{M_t, \mathbf{F}_t;\ t \in I\}$ is a regular right continuous martingale such that $M_t = E(F|\mathbf{F}_t)$ for some $F = F(\omega) \in L^1(\Omega)$, then for any stopping time τ with $P(\tau < \infty) = 1$ we have*

$$M_\tau = E(F|\mathbf{F}_\tau), \quad \text{a.e. } P.$$

PROOF. Define a stopping time τ_n by $\tau_n(\omega) = k/2^n$ if $\tau(\omega) \in [(k-1)/2^n, k/2^n)$. Then, $\tau_n(\omega) \downarrow \tau(\omega)$ as $n \to \infty$. Since for any $A \in \mathbf{F}_{\tau_n}$,

$$\int_{A \cap \{\tau_n = k/2^n\}} M_{k/2^n} P(d\omega) = \int_{A \cap \{\tau_n = k/2^n\}} F(\omega)\, P(d\omega), \qquad k = 1, 2, \ldots,$$

we have

$$\int_A M_{\tau_n} P(d\omega) = \sum_{k=1}^{\infty} \int_{A \cap \{\tau_n = k/2^n\}} M_{k/2^n} P(d\omega) = \int_A F(\omega)\, P(d\omega),$$

which means $M_{\tau_n} = E(F|\mathbf{F}_{\tau_n})$. Hence $\{M_{\tau_n};\ n = 1, 2, \ldots\}$ is a system of uniformly integrable random variables. Using the right continuity of M, we have for any $A \in \mathbf{F}_\tau$ $(\subset \mathbf{F}_{\tau_n},\ n = 1, 2, \ldots)$,

$$\int_A M_\tau P(d\omega) = \lim_{n \to \infty} \int_A M_{\tau_n} P(d\omega) = \int_A F(\omega)\, P(d\omega).$$

This means that $M_\tau = E(F|\mathbf{F}_\tau)$, a.e. P. □

The following theorem is directly derived from Lemma A.2.

THEOREM A.5. *Let M be a regular martingale and let σ and τ be two stopping times such that $\sigma \leq \tau$, a.e. P. Then*

$$M_\sigma = E(M_\tau|\mathbf{B}_\sigma), \quad \text{a.e. } P.$$

COROLLARY. *Let $M = \{M_t, \mathbf{F}_t;\ t \in I\}$ be a right continuous martingale. Then for any stopping time τ, $M^\tau = \{M_{t \wedge \tau}, \mathbf{F}_t;\ t \in I\}$ is again a right continuous martingale.*

PROOF. For a stopping time $t \wedge \tau$, Theorem A.5. can be applied. In fact, for fixed $t_0 \in I$, the process $\{M_t, \mathbf{F}_t;\ t \in I_0\}$, $I_0 \equiv \{t;\ t \leq t_0\} \cap I$, is a regular martingale. □

2°. Stochastic (Itô) integrals. In this subsection, we define the Itô type stochastic integral based on a Brownian motion. Let $B = \{B(t), \mathbf{F}_t;\ t \in I\}$, $I = [0, \infty)$, be a Brownian motion, where $\{\mathbf{F}_t;\ t \in I\}$ is an increasing system of sub-σ-fields. See the remark in §2.5 for the notation $\{B(t), \mathbf{F}_t;\ t \in I\}$.

Let $\mathscr{F}$ be a set of stochastic processes $f = \{f(s, \omega);\ s \in I\}$, $I = [0, \infty)$, satisfying

(*f*.1) $f(t, \omega)$ is (t, ω)-measurable,
(*f*.2) $f(t, \omega)$ is $\mathbf{F}_t$-measurable for each $t \in I$,
(*f*.3) $E(\int_0^t f(s, u)^2\, ds) < \infty$, $t \in I$.

The *stochastic integral* $\int_0^t f(s, \omega)\, dB(s)$ for $f \in \mathscr{F}$ is defined in the following steps.

The first step. If $f(s, \omega)$ is $\mathbf{F}_0$-measurable for all $s \in I$, the integral is defined as the Wiener integral. In fact, $f(s, \omega)$ is independent of B, so f is like a member of $L^2([0, t])$ for any $t \in I$.

The second step. If there exists an $h > 0$ and if $f(s, \omega)$ is $\mathbf{F}_{s-h}$-measurable for each $s \in I$, where we set $\mathbf{F}_t = \mathbf{F}_0$, $t \leq 0$, then the stochastic integral is defined by

$$\int_0^t f(s, \omega)\, dB(s) \equiv \sum_{i=1}^{\infty} \int_{(i-1)/n)\wedge t}^{(i/n)\wedge t} f(s, \omega)\, dB(s), \tag{A.10}$$

where n is chosen so that $1/n < h$. Note that the right-hand side of (A.10) makes sense, because the $f(s, \omega)$, $s \in [(i-1)/n, i/n)$, are independent of $B(u) - B((i-1)/n)$, for $u \geq (i-1)/n$, so each term on the right-hand side is regarded as an Wiener integral. Note that the sum is independent of $n\ (> 1/h)$.

Before going to the third step, we provide the following

LEMMA A.4. *The mean* $E(\int_0^t f(s, \omega)\, dB(s))$ *is* 0, *and the variance* $E(\int_0^t f(s, \omega)\, dB(s))^2$ *is equal to* $\int_0^t E(f(s, \omega)^2)\, ds$ *for any* $f \in \mathscr{F}$ *provided that* $f(s, \omega)$ *is* $\mathbf{F}_{s-h}$*-measurable for some* $h > 0$.

PROOF. The proof is easy once we check a trivial fact

$$E\left(\int_{(i-1)/n}^{i/n} f(s, \omega)\, dB(s) \int_{(j-1)/n}^{j/n} f(s, \omega)\, dB(s)\right) = \delta_{ij} \int_{(i-1)/n}^{i/n} E(f(s, \omega)^2)\, ds,$$

where δ_{ij} is Kronecker's delta. □

The third step. Suppose $f = \{f(s, \omega),\ s \in I\} \in \mathscr{F}$ is bounded: $|f(s, \omega)| \leq M$. For such an f, we know that

$$\int_0^t (f(s, \omega) - f(s-h, \omega))^2\, ds \to 0 \quad \text{as } h \to 0 \quad \text{a.e. } P,$$

where we regard $f(s, \omega)$ as 0 for $s < 0$. Using the bounded convergence theorem, we obtain

$$E\left(\int_0^t (f(s, \omega) - f(s-h, \omega))^2 \, ds\right) \to 0 \quad \text{as } h \to 0, \tag{A.11}$$

for any $t \in I$. This means that

$$\begin{aligned} E&\left(\left|\int_0^t f(s-h, \omega)\, dB(s) - \int_0^t f(s-h', \omega)\, dB(s)\right|^2\right) \\ &= E\left(\left|\int_0^t f(s-h, \omega) - f(s-h', \omega)\, dB(s)\right|^2\right) \\ &= \int_0^t E(|f(s-h, \omega) - f(s-h', \omega)|^2)\, ds \to 0 \quad \text{as } h, h' \to 0, \end{aligned}$$

by using Lemma A.3. Thus, there exists a limit

$$I_t(f) = \lim_{h \to 0} \int_0^t f(s-h, \omega)\, dB(s)$$

in the Hilbert space $L^2(\Omega, \mathbf{F}, P)$. The limit $I_t(f)$ is denoted by $\int_0^t f(s)\, dB(s)$ and is called the stochastic integral for bounded f. Lemma A.5 follows easily from the definition of stochastic integral.

LEMMA A.5. *The stochastic integral for bounded $f \in \mathcal{F}$ has the same properties as in Lemma A.4. That is,*

$$E\left(\int_0^t f(s, \omega)\, dB(s)\right) = 0$$

and

$$E\left(\left|\int_0^t f(s, \omega)\, dB(s)\right|^2\right) = \int_0^t E(f(s, \omega)^2)\, ds.$$

The fourth step. We now define a stochastic integral for general f in $\mathcal{F}$. Set $f_N(s, \omega) = f(s, \omega) \wedge N$. For bounded f_N we define a sequence of stochastic integrals $\int_0^t f_N(s, \omega)\, dB(s)$, $N = 1, 2, \ldots$, by the method in the third step. Then Lemma A.4. guarantees

$$\begin{aligned} E&\left(\left|\int_0^t f_N(s, \omega)\, dB(s) - \int_0^t f_{N'}(s, \omega)\, dB(s)\right|^2\right)^2 \\ &= \int_0^t E(|f_N(s, \omega) - f_{N'}(s, \omega)|^2)\, ds \to 0 \quad \text{as } N, N' \to \infty. \end{aligned}$$

Thus, there exists a limit $I_t(f) = \lim_{N \to \infty} \int_0^t f_N(s, \omega)\, dB(s)$ in $L^2(\Omega, \mathbf{B})$. $I_t(f)$ is the stochastic integral of $f = \{f(s, \omega);\ s \in I\}$, denoted by $\int_0^t f(s, \omega)\, dB(s)$.

REMARK. The stochastic integral $\int_u^t f(s, \omega)\, dB(s)$ is defined as

$$\int_0^t f(s, \omega)\, dB(s) - \int_0^u f(s, \omega)\, dB(s).$$

The following theorem lists properties of stochastic integrals.

THEOREM A.6. (1) $\int_u^t (\alpha f_1 + \beta f_2)\, dB = \alpha \int_u^t f_1\, dB + \beta \int_u^t f_2\, dB$ *a.e.* P, *for any* $u, t \in I$, $f_1, f_2 \in \mathscr{F}$, α, β; *constants.*

(2) $E((\int_u^t f(s, \omega) dB(s))^2) = \int_u^t E(f(s, \omega)^2)\, ds$, $u, t \in I$, $f \in \mathscr{F}$,

(3) *For* $f \in \mathscr{F}$, $M = \{M_t, \mathbf{F}_t;\ t \in [0, \infty)\}$ *with* $M_t = \int_0^t f(s, \omega)\, dB(s)$, *is a martingale. If, in addition,* $E(\int_0^\infty f(s, \omega)^2\, ds) < \infty$, *then* M *is regular.*

PROOF. (1) and (2) are obvious from the definition. The first part of (3) follows from a simple calculation. If, in particular $E(\int_0^\infty f(s, \omega)^2\, ds) < \infty$, we can define a stochastic integral $F(\omega) = \int_0^\infty f(s, \omega)\, dB(s)$ and obtain $M_t = E(F|\mathbf{F}_t)$. □

Next we study the continuity of a martingale given by stochastic integrals.

THEOREM A.7. *A martingale* $M = \{M_t, \mathbf{F}_t;\ t \in [0, \infty)\}$, *with* $M_t = \int_0^t f(s, \omega)\, dB(s)$, $f \in \mathscr{F}$, *is continuous.*

PROOF. *The first step.* Let M_t be written in the form $M_t = \int_0^t f(s, \omega)\, dB(s)$ with $\mathbf{F}_0$-measurable $f(s, \omega)$, $s \in I$. By the well-known property of the Wiener integral, it is easy to prove M_t is continuous in t with probability one.

The second step. If $f(s, \omega)$ is $\mathbf{F}_{s-h}$-measurable, then the continuity of M_t is trivial from the definition (A.10) of the stochastic integral, because the stochastic integral is a sum of (local) Wiener integrals.

The third step. For a general $f = f(s, \omega) \in \mathscr{F}$, we can choose a sequence $\{f_n(s, \omega);\ n = 1, 2, \ldots\}$ of $\mathbf{F}_{s-1/n}$-measurable functions for each $s \in I$, satisfying $E(\int_0^t (f_n(s, \omega) - f(s, \omega))^2\, ds) \to 0$, as $n \to \infty$, for any t (cf. third and fourth steps in the definition of stochastic integral).

Set $M_t^{(n)} = \int_0^t f_n(s, \omega)\, dB(s)$, then $M - M^{(n)} = \{M_t - M_t^{(n)}, \mathbf{F}_t;\ t \in [0, \infty)\}$ is a martingale. Hence, $(M - M^{(n)})^2 = \{(M_t - M_t^{(n)})^2, \mathbf{F}_t;\ t \in [0, \infty)\}$ is a submartingale. Applying Doob's inequality (Theorem A.4), we obtain

$$P\left(\sup_{u \le t} |M_u - M_u^{(n)}| \ge \lambda\right) \le \frac{1}{\lambda^2} E((M_t - M_t^{(n)})^2)$$
$$= \frac{1}{\lambda^2} \int_0^t E((f(s, \omega) - f_n(s, \omega))^2)\, ds \to 0 \quad \text{as } n \to \infty.$$

This means that $\sup_{u \le t} |M_u - M_u^{(n)}|$ converges to zero in probability; therefore, there exists a subsequence $\sup_{u \le t} |M_u - M_u^{(n_k)}|$ converges to zero with probability one. As each $M_u^{(n_k)}$ is continuous, the uniform limit M_u is also continuous. Since $t \in [0, \infty)$ is fixed arbitrarily, the continuity of the martingale M is proved. □

REMARK. In general, if f_n and $f \in \mathscr{F}$ satisfy the condition

$$\lim_{n\to\infty} \int_0^t E((f_n(s, \omega) - f(s, \omega))^2)\, ds = 0, \qquad t \in [0, \infty),$$

then $M_t^{(n)} = \int_0^t f_n(s, \omega)\, dB(s)$ converges uniformly in the wide sense to $M_t = \int_0^t f(s, \omega)\, dB(s)$, in probability.

From Theorem A.7 we see that stochastic integrals are defined for $f(s, \omega)$ in a much wider class than $\mathscr{F}$. Let $\mathscr{G}$ be a class of stochastic processes f satisfying $(f.1)$, $(f.2)$ and

$(f'.3)$ $\int_0^t f(s, \omega)^2\, ds < \infty$ a.e. P, for any $t \in I$.

DEFINITION A.7. For $f \in \mathscr{G}$ a stochastic integral is defined to be

$$\int_0^t f(s, \omega)\, dB(s) = \lim_{N\to\infty} \int_0^{t\wedge\tau_N} f(s, \omega)\, dB(s)$$
$$\left(= \lim_{N\to\infty} \int_0^t \chi_{[0, \tau_N]}(s) f(s, \omega)\, dB(s)\right), \tag{A.11}$$

where $\tau_N = \inf\{t;\ \int_0^t f(s, \omega)^2\, ds \geq N\}$.

REMARK. (1) τ_N is a stopping time for each N.

(2) Because τ_N is larger than t for sufficiently large N with probability one, the limit in the right-hand side of (A.11) exists.

(3) For $f \in \mathscr{F}$, the above definitions of a stochastic integral is in agreement with the one given before. Indeed, the right-hand side of (A.11) is an approximation of the stochastic integral of $f \in \mathscr{F}$.

(4) For $f \in \mathscr{G}$, $M_t = \int_0^t f(s, \omega)\, dB(s)$ may not be a martingale, in fact, it may not have mean value. It is approximated by a sequence of square integrable martingales $M_t^N = \int_0^{t\wedge\tau_N} f(s, \omega)\, dB$.

THEOREM A.8. *If a sequence of stochastic process* $\{f_n;\ n = 1, 2, \ldots\}$ *in* $\mathscr{G}$ *satisfies*

$$\int_0^t (f(s, \omega) - f_n(s, \omega))^2\, ds \to 0 \quad \text{as } n \to \infty, \tag{A.12}$$

in probability, then $M_t^{(n)} = \int_0^t f_n(s, \omega)\, dB(s)$ *converges uniformly in the wide sense to* $M_t = \int_0^t f(s, \omega)\, dB(s)$, *in probability*:

$$P\left(\sup_{s\leq t} |M_t^{(n)} - M_t| > \lambda\right) \to 0, \quad \text{for any } \lambda > 0,\ t \in I.$$

PROOF. For a sequence of stopping times

$$\left\{\tau_{n,N} = \inf\left\{t;\ \int_0^t f_n(s, \omega)^2\, ds \geq N\right\};\ n = 1, 2, \ldots\right\},$$

$$E\left(\int_0^t (\chi_{[0, \tau_{n,N})}(s) f_n(s, \omega) - \chi_{[0, \tau_N)}(s) f(s, \omega))^2\, ds\right) \to 0, \qquad \text{as } n \to \infty, \tag{A.13}$$

by (A.12). Therefore, $\int_0^{t\wedge\tau_{n,N}} f_n\,dB$ converges uniformly in the wide sense to $\int_0^{t\wedge\tau_N} f\,dB$, in probability. Note the inequality

$$P\left(\sup_{s\le t}|M_s^{(n)} - M_s| > \lambda\right)$$
$$\le P\left(\sup_{s\le t}|M_s^{(n)} - M_s| > \lambda,\ \tau_{n,N}, \tau_N \ge t\right) + P(\tau_{n,N} < t) + P(\tau_N < t).$$

By (A.13) the first term on the right-hand side converges to zero as $n \to \infty$. The second and third terms are equal to $P(\int_0^t f_n^2\,ds > N)$ and $P(\int_0^t f^2\,ds > N)$, respectively. For a sufficiently large N, we get $P(\int_0^t f_n^2\,ds > N) < \varepsilon$, $n = 1, 2, \ldots$, and $P(\int_0^t f^2\,ds > N) < \varepsilon$. Hence the result follows. □

The remarks below are easy consequences of this Theorem.

REMARK 1. The stochastic process defined by $\int_0^t f\,dB$, $f \in \mathscr{G}$, is continuous in t with probability one.

REMARK 2. In Definition A.7, we took, for $f \in \mathscr{G}$, a sequence $\{f_N = f\cdot\chi_{[0,N]} \in \mathscr{F};\ N = 1, 2, \ldots\}$ to define the integral $\int_0^t f\,dB$. We see, from Theorem A.8, that $\int_0^t f_n\,dB$ converges to the same limit $\int_0^t f\,dB$ for any sequence $\{f_n;\ n = 1, 2, \ldots\}$ such that $\int_0^t (f - f_n)^2\,ds \to 0$, in probability. As a result, we prove that $\int_0^t f\,dB$, $f \in \mathscr{G}$, is well defined.

For the rest of this section, we focus on a relation between Wiener integrals and stochastic (Itô) integrals. Let $f = \{f(s, \omega);\ s \in I\}$ be in $\mathscr{F}$. If $f(s, \omega)$ is $\mathbf{B}_s(B)$-measurable for each $s \in I$, then it has the following expansion in terms of multiple Wiener integrals:

$$f(s, \omega) = \sum_{n=0}^{\infty} \int_0^s \int_0^{t_1} \cdots \int_0^{t_{n-1}} k_n(s, t_1, \ldots, t_n)\,dB(t_n)\cdots dB(t_1). \quad \text{(A.14)}$$

The kernels $k_n(s, t_1, \ldots, t_n)$ are supposed to be jointly measurable. We pause here to note that the restriction of the basic measure space $(\Omega, \mathbf{B}, P) = (C, \mathbf{C}, \mu_W)$ does not disturb the results in Theorems A.1, A.2, and A.3.

PROPOSITION A.2. (1) *If $f \in \mathscr{F}$ can be written in the form* (A.14), *then*

$$\int_0^t f(s, \omega)\,dB(s)$$
$$= \sum_{n=0}^{\infty} \int_0^t \int_0^s \int_0^{t_1} \cdots \int_0^{t_{n-1}} k_n(s, t_1, \ldots, t_n)\,dB(t_n)\cdots dB(t_1)\,dB(s). \quad \text{(A.15)}$$

The left-hand side is a stochastic integral, while the right-hand side is a multiple Wiener integral.

(2) *If $f(s, \omega) \in \mathscr{H}_n(s)$, $s \in I$, then $\int_0^t f(s, \omega)\,dB(s) \in \mathscr{H}_{n+1}(t)$, $t \in I$.*

(3) *A martingale $M = \{M_t, \mathbf{B}_t(B);\ t \in [0, \infty)\}$ with $E(M_t) = 0$ and with $E(M_t^2) < \infty$, $t \in I$, can be written in the form*

$$M_t = \int_0^t f(s, \omega)\,dB(s),$$

for a suitable $f \in \mathcal{F}$ such that $f(s, \omega)$ is $\mathbf{B}_s(B)$-measurable.

PROOF. The assertions (1) and (2) are rather trivial from the definitions of both integrals. Statement (3) is clear from the definition of the Wiener expansion. In fact, for a fixed $t_0 \in [0, \infty)$, M_{t_0} has an expansion of the form

$$M_{t_0} = \sum_{n=1}^{\infty} \int_0^{t_0} \int_0^{t_1} \cdots \int_0^{t_{n-1}} k_n(t_0; t_1, \dots, t_n)\, dB(t_n) \cdots dB(t_1).$$

Then we have

$$\begin{aligned} M_t &= E(M_{t_0} | \mathbf{B}_t(B)) \\ &= \sum_{n=0}^{\infty} \int_0^{t} \int_0^{t_1} \cdots \int_0^{t_{n-1}} k_n(t_0; t_1, \dots, t_n)\, dB(t_n) \cdots dB(t_1), \end{aligned}$$

for $t \le t_0$. The uniqueness of the expansion proves that each kernel $k_n(t_0, t_1, \dots, t_n)$ does not depend on t_0:

$$k_n(t_0, t_1, \dots, t_n) = k_n(t_1, \dots, t_n).$$

Hence, we define $f = \{f(s, \omega);\ s \in I\}$ by

$$f(s, \omega) = k_1(s) + \sum_{n=2}^{\infty} \int_0^{s} \int_0^{t_2} \cdots \int_0^{t_{n-1}} k_n(s, t_2, \dots, t_n)\, dB(t_n) \cdots dB(t_2).$$

Then, the process f belongs to $\mathcal{F}$, and further $f(s, \omega)$ is $\mathbf{B}_s(B)$-measurable for each s. The expression $M_t = \int_0^t f\, dB$ follows from assertion (1). □

Proposition A.2. (3) can now be generalized to

THEOREM A.9. *A martingale $M = \{M_t, \mathbf{B}_t(B);\ t \in [0, \infty)\}$ on the Wiener space is continuous, and there exists a process $f = \{f(s, \omega);\ s \in I\}$ such that $f(s, \omega)$ is $\mathbf{B}_s(B)$-measurable for each s, and such that M can be written in the form*

$$M_t = \int_0^t f(s, \omega)\, dB(s), \qquad t \in [0, \infty). \tag{A.16}$$

PROOF. First, we prove continuity. Take $t_0 \in [0, \infty)$ and N. Set

$$M_{t_0}^N = \begin{cases} M_{t_0}, & \text{if } |M_{t_0}| \le N, \\ N, & \text{if } M_{t_0} > N, \\ -N, & \text{if } M_{t_0} < -N. \end{cases}$$

Then we have a martingale $M^N = \{M_t^N = E(M_{t_0}^N | \mathbf{B}_t(B)), \mathbf{B}_t(B);\ t \in [0, t_0]\}$. The random variable M_t^N is square integrable for each $t \in [0, t_0]$ (in fact it is bounded: $|M_t| \le N$), and it can be written as

$$M_t^N = \int_0^t f_N(s, \omega)\, dB(s),$$

from which it is seen that M^N is continuous. Since $|M_t^N - M_t|$ is a submartingale, Doob's inequality proves

$$P\left(\sup_{0\le t\le t_0} |M_t^N - M_t| > \lambda\right) \le \frac{1}{\lambda}E(|M_{t_0}^N - M_{t_0}|) \to 0 \quad \text{as } N \to \infty,$$

This implies that M_t^N converges uniformly to M_t in probability. Hence, we see that M_t is continuous on the interval $[0, t_0]$. Therefore, M is a continuous martingale.

Next we derive (A.16). Define stopping times τ_k; $k = 1, 2, \dots$, by $\tau_k = \inf\{t;\ |M_t| > k\}$ (where $\inf\{\phi\} = \infty$). From the corollary to Theorem A.5, $M^k = \{M_t^k = M_{t\wedge\tau_k}, \mathbf{B}_t(B);\ t \in [0, \infty)\}$ is a martingale. Since $E(M_{t\wedge\tau_k}^2) \le K^2 < \infty$, there exists an $f_k = \{f_k(t, \omega);\ t \in I\} \in \mathscr{F}$ such that

$$M_{t\wedge\tau_k} = \int_0^t f_k(s, \omega)\, dB(s),$$

according to Theorem A.6(3). Since

$$E\left(\int_0^{t_0\wedge\tau_k} |f_k - f_{k'}|^2\, ds\right) = E(|M_{t_0\wedge\tau_k} - M_{t_0\wedge\tau_{k'}\wedge\tau_k}|^2) = 0, \qquad k' > k,$$

we know that for almost all $\omega \in \Omega$,

$$f_k(t, \omega) = f_{k'}(t, \omega), \qquad k' > k, \quad t \le t_0$$

for k sufficiently large. Therefore, we can set $f(t, \omega) = f_k(t, \omega)$. Since $\int_0^{t_0} f^2\, ds < \infty$, a.e. P, we can define $\int_0^t f\, dB$ for $t \le t_0$. If $\tau_k > t_0$, we have $\int_0^t f\, dB = \int_0^t f_k\, dB = M_{t\wedge\tau_k} = M_t$, $t \le t_0$. This means that $M_t = \int_0^t f(s, \omega)\, dB(s)$, $t \le t_0$. Since t_0 is arbitrary, the expression (A.16) is proved. □

§3. Itô's formula and Girsanov's theorem

A smooth function of a stochastic integral is again expressed as a sum of a stochastic integral. We are interested in its variation. The result is similar, in a sense, to the formula for the variation of an ordinary function in elementary calculus, however the crucial difference in Itô's formula appears when we are concerned with functions of a stochastic integral; there one is asked to use the second derivative of a function to express the differential. The reason is that the Brownian paths are not of bounded variation, but $(dB)^2 \doteqdot dt$ as we have seen in §2.5. We are now ready to state

THEOREM A.10 (Itô's formula). *Let a function $u = u(t, x)$ be of class C^1 in t and of class C^2 in x: $u, \partial u/\partial t, \partial u/\partial x$, and $\partial^2 u/\partial x^2$ are all jointly continuous. Suppose that $f = \{f(t, \omega);\ t \in I\}$ is in $\mathscr{G}$ and that $g = \{g(t, \omega);\ t \in I\}$ satisfies $(f.1)$, $(f.2')$, and* (g.3) $\int_0^t |g(s, \omega)|\,ds < \infty$ *a.e.* P, $t \in [0, \infty)$. *Then for the process* $\alpha = \{\alpha(t);\ t \in I\}$ *given by*

$$\alpha(t) = \int_0^t f(s, \omega)\,dB(s) + \int_0^t g(s, \omega)\,ds,$$

we have

$$\begin{aligned} u(t_1, \alpha(t_1)) &- u(t_0, \alpha(t_0)) \\ &= \int_{t_0}^{t_1} \frac{\partial u}{\partial x}(s, \alpha(t))\,d\alpha(t) \\ &+ \int_{t_0}^{t_1} \left(\frac{\partial u}{\partial t}(t, \alpha(t)) + \frac{1}{2}\frac{\partial^2 u}{\partial x^2}(t, \alpha(t)) f(t, \omega)^2 \right) dt, \quad t_1 > t_0. \end{aligned} \tag{A.17}$$

Outline of the Proof. *The first step.* Suppose that both $f(t, \omega)$ and $g(t, \omega)$ are $\mathbf{F}_0$-measurable for each $t \in I$. Take a division Δ: $t_0 = \tau_0 < \tau_1 < \cdots < \tau_N = t_1$. Then, from the Taylor expansion we have

$$\begin{aligned} u(t_1, \alpha(t_1)) &- u(t_0, \alpha(t_0)) \\ &= \sum_{k=1}^{N} (u(\tau_k, \alpha(\tau_k)) - u(\tau_{k-1}, \alpha(\tau_k))) \\ &+ \sum_{k=1}^{N} (u(\tau_{k-1}, \alpha(\tau_k)) - u(\tau_{k-1}, \alpha(\tau_{k-1}))) \\ &= \sum_{k=1}^{N} \left(\frac{\partial u}{\partial t}(\tau_{k-1}, \alpha(\tau_k)) + \varepsilon_1(\tau_{k-1}, \tau_k, \alpha(\tau_k)) \right)(\tau_k - \tau_{k-1}) \\ &+ \sum_{k=1}^{N} \left\{ \frac{\partial u}{\partial x}(\tau_{k-1}, \alpha(\tau_{k-1}))(\alpha(\tau_k) - \alpha(\tau_{k-1})) \right. \\ &\quad + \left(\frac{1}{2}\frac{\partial^2 u}{\partial x^2}(\tau_{k-1}, \alpha(\tau_{k-1})) + \varepsilon_2(\tau_{k-1}, \alpha(\tau_{k-1}), \alpha(\tau_k)) \right) \\ &\quad \left. (\alpha(\tau_k) - \alpha(\tau_{k-1}))^2 \right\}. \end{aligned}$$

Now let $|\Delta| = \max\{|\tau_k - \tau_{k-1}|;\ k = 1, 2, \ldots, N\}$ tend to zero. Then we will prove that the right-hand side of the above equality converges in probability to

$$\int_{t_0}^{t_1} \frac{\partial u}{\partial t}(t, \alpha(t))\,dt + \int_{t_0}^{t_1} \frac{\partial u}{\partial x}(t, \alpha(t))\,d\alpha(t) + \frac{1}{2}\int_{t_0}^{t_1} \frac{\partial^2 u}{\partial x^2}(t, \alpha(t)) f(t, \omega)^2\,dt.$$

As for the first term, the equality

$$u(\tau_k, \alpha(\tau_k)) - u(\tau_{k-1}, \alpha(\tau_k))$$
$$= \frac{\partial u}{\partial t}(\tau_{k-1}, \alpha(\tau_k))(\tau_k - \tau_{k-1})$$
$$+ \left(\frac{\partial u}{\partial t}(\tau_{k-1} + \theta(\tau_k - \tau_{k-1}), \alpha(\tau_k)) - \frac{\partial u}{\partial t}(\tau_{k-1}, \alpha(\tau_k))\right)(\tau_k - \tau_{k-1}), \quad 0 < \theta < 1,$$
$$= \frac{\partial u}{\partial t}(\tau_{k-1}, \alpha(\tau_k))(\tau_k - \tau_{k-1}) + \varepsilon_1(\tau_{k-1}, \tau_k, \alpha(\tau_k))(\tau_k - \tau_{k-1})$$

yields

$$\sum_{k=1}^{N} \frac{\partial u}{\partial t}(\tau_{k-1}, \alpha(\tau_k))(\tau_k - \tau_{k-1}) \to \int_{t_0}^{t_1} \frac{\partial u}{\partial t}(t, \alpha(t))\, dt$$

and

$$\sum_{k=1}^{N} \varepsilon_1(\tau_{k-1}, \tau_k, \alpha(\tau_k))(\tau_k - \tau_{k-1}) \to 0$$

in probability as $n \to \infty$, since $\partial u(t, \alpha(t))/\partial t$ is uniformly continuous with probability one by the continuity of $\alpha(t)$ and $\partial u(t, x)/\partial t$. For the second term, the uniform continuity of $\partial u(t, \alpha(t))/\partial x$ shows that

$$\sum_{k=1}^{N} \frac{\partial u}{\partial x}(\tau_{k-1}, \alpha(\tau_{k-1}))(\alpha(\tau_k) - \alpha(\tau_{k-1})) \to \int_{t_0}^{t_1} \frac{\partial u}{\partial x}(t, \alpha(t))\, d\alpha(t)$$

in probability, as $N \to \infty$. Now consider the term

$$(\alpha(\tau_k) - \alpha(\tau_{k-1}))^2$$
$$= \left(\int_{\tau_{k-1}}^{\tau_k} f\, dB\right)^2 + 2\left(\int_{\tau_{k-1}}^{\tau_k} f\, dB\right)\left(\int_{\tau_{k-1}}^{\tau_k} g\, ds\right) + \left(\int_{\tau_{k-1}}^{\tau_k} g\, ds\right)^2.$$

By the corollary to Lemma A.1, we have

$$\left(\int_{\tau_{k-1}}^{\tau_k} f\, dB\right)^2 = \int_{\tau_{k-1}}^{\tau_k} f^2\, ds + 2\int_{\tau_{k-1}}^{\tau_k} \left(\int_{\tau_{k-1}}^{t} f\, dB\right) f\, dB,$$

since $H_2(t, x) = \frac{1}{2}(x^2 - t)$. Therefore,

$$\sum_{k=1}^{N} \frac{\partial^2 u}{\partial x^2}(\tau_{k-1}, \alpha(\tau_{k-1}))\left(\int_{\tau_{k-1}}^{\tau_k} f\, dB\right)^2 \to \int_{t_0}^{t_1} \frac{\partial^2 u}{\partial x^2}(t, \alpha(t)) f(t, \omega)^2\, dt,$$

in probability, as $n \to \infty$, because $\partial^2 u(t, \alpha(t))/\partial x^2$ is continuous and $\int_{t_0}^{t_1} \varphi_N(t, \omega)^2\, dt \to 0$, in probability, where

$$\varphi_N(t, \omega) = \frac{\partial^2 u}{\partial x^2}(\tau_{k-1}, \alpha(\tau_{k-1}))\left(\int_{\tau_{k-1}}^{t} f\, dB\right) f(t, \omega), \quad \text{for } t \in [\tau_{k-1}, \tau_k),$$

$k = 1, 2, \ldots, N$. (see Theorem A.8.). It is easy to see that all the remaining summations converge to zero:

$$\sum_{k=1}^{N}\left\{\frac{\partial^2 u}{\partial x^2}(\tau_{k-1}, \alpha(\tau_{k-1}))\left(2\left(\int_{\tau_{k-1}}^{\tau_k} f\,dB\right)\left(\int_{\tau_{k-1}}^{\tau_k} g\,ds\right)+\left(\int_{\tau_{k-1}}^{\tau_k} g\,ds\right)^2\right)\right.$$
$$\left.+\varepsilon_2(\tau_{k-1}, \alpha(\tau_{k-1}), \alpha(\tau_k))(\alpha(\tau_k)-\alpha(\tau_{k-1}))^2\right\} \to 0,$$

in probability, as $n \to \infty$.

The second step. For $f = \{f(s, \omega);\ s \in I\}$ and $g = \{g(s, \omega);\ s \in I\}$ such that both $f(s, \omega)$ and $g(s, \omega)$ are $\mathbf{F}_{s-h}$-measurable for some $h > 0$, (A.17) can still be proved by repeating the arguments in the first step.

The third step. For a general $f = \{f(s, \omega);\ s \in I\}$ and $g = \{g(s, \omega);\ s \in I\}$, define $f_n = \{f_n(s, \omega);\ s \in I\}$ and $g_n = \{g(s, \omega);\ s \in I\}$, by $f_n(s, \omega) = f(s - 1/n, \omega)$ and $g_n(s, \omega) = g(s - 1/n, \omega)$, for $s \geq 1/n$, and $f_n(s, \omega) = g_n(s, \omega) = 0$, for $s < 1/n$, respectively. Noting that

$$\int_0^t (f_n(s, \omega) - f(s, \omega))^2\,ds \to 0 \quad \text{and} \quad \int_0^t |g_n(s, \omega) - g(s, \omega)|\,ds \to 0,$$

in probability, as $n \to \infty$, $t \in I$, we can see that $\alpha_n(t) = \int_0^t f_n\,dB + \int_0^t g_n\,dt$ converges to $\alpha(t)$, in probability. Hence, from the continuity of $\partial u/\partial t$, $\partial u/\partial x$, and $\partial^2 u/\partial x^2$ we have that

$$\int_{t_0}^{t_1} \frac{\partial u}{\partial x}(s, \alpha_n(s))\,d\alpha_n(s)$$
$$+\int_{t_0}^{t_1}\left(\frac{\partial u}{\partial t}(s, \alpha_n(s)) + \frac{1}{2}\frac{\partial^2 u}{\partial x^2}(s, \alpha_n(s))f_n(s, \omega)^2\right)ds$$

converges in probability to the right-hand side of (A.17) (by Theorem A.8).
□

REMARK. The method in the proof of (A.17) is analogous to the proof of Lemma A.1. The formulae of Hermite polynomials (Proposition A.1(ii)) is a particular case of Itô's formula.

EXAMPLE.

$$M_{s,t} \equiv \exp\left(\int_s^t f(u, \omega)\,dB(u) - \frac{1}{2}\int_s^t f(u, \omega)^2\,du\right)$$
$$= 1 + \int_s^t M_{s,u} f(u, \omega)\,dB(u),$$

where $f \in \mathscr{G}$.

As an application of Itô's formula, we prove Girsanov's theorem, which is important for the study of equivalence problems (§6.4).

Theorem A.11 (I. V. Girsanov). *Let a stochastic process* $f = \{f(s, \omega);\ s \in I\}$ *in* $\mathcal{G}$ *be given. If* $M_{s,u} \equiv \exp(\int_s^t f\,dB(v) - \frac{1}{2}\int_s^t f^2\,dv)$, $u > s$, *satisfies the condition*

$$E(M_{0,T}) = 1 \quad \textit{for a fixed } T > 0, \tag{A.18}$$

then, the process $X = \{X(t), \mathbf{F}_t;\ t \in [0, T]\}$ *defined by*

$$X(t) = B(t) - \int_0^t f(s, \omega)\,ds$$

is a Brownian motion on the new probability space $(\Omega, \mathbf{B}, \widetilde{P})$ *formed by setting* $\widetilde{P}(d\omega) = M_{0,T}P(d\omega)$.

Proof. It is easily seen that $\{M_{0,t}, \mathbf{F}_t;\ t \in [0, T]\}$ is a martingale by assumption (A.18): $E(M_{0,T}|\mathbf{F}_t) = M_{0,t}$ a.e. P, $t \in [0, T]$. Compute

$$\alpha_{u,t}(\lambda) \equiv \widetilde{E}\left(\exp(i\lambda(X(t) - X(u))|\mathbf{F}_u\right) \tag{A.19}$$

where $\widetilde{E}(\ |\ )$ denotes the conditional expectation on the probability space $(\Omega, \mathbf{B}, \widetilde{P})$. As $\widetilde{E}(F|\mathbf{F}_u) = E(FM_{u,T}|\mathbf{F}_u)$, a.e. P and a.e. $\widetilde{P}$, too, for any bounded random variable $F = F(\omega)$, the right-hand side of (A.19) is equal to

$$E\left(\exp\left(i\lambda(X(t) - X(u)) + \int_u^T f\,dB(s) - \frac{1}{2}\int_u^T f^2\,ds\right)\Big|\,\mathbf{F}_u\right)$$
$$= E\left(\exp\left(\int_u^t (i\lambda + f)\,dB(s) - \int_u^t (i\lambda f + f^2/2)\,ds\right)\Big|\,\mathbf{F}_u\right)$$

a.e. P and a.e. $\widetilde{P}$. Applying Itô's formula, the last expectation above is

$$E\left(1 + \int_u^t \exp\left(\int_u^s (i\lambda + f)\,dB(v) - \int_u^s (i\lambda f + f^2/2)\,dv\right)(i\lambda + f)\,dB(s)\right.$$
$$\left. - \int_u^t \exp\left(\int_u^s (i\lambda + f)\,dB(v) - \int_u^s (i\lambda f + f^2/2)\,dv\right)\frac{-\lambda^2}{2}\,ds\,\Big|\,\mathbf{F}_u\right)$$
$$= 1 - \frac{\lambda^2}{2}\int_u^t \alpha_{u,s}(\lambda)\,ds.$$

In fact, $E(\int_u^t \exp(\cdots)(i\lambda + f)\,dB(s)|\mathbf{F}_u) = 0$, since $E(M_{u,t}|\mathbf{F}_u) = 1$. The integral equation

$$\alpha_{u,t}(\lambda) = 1 - \frac{\lambda^2}{2}\int_u^t \alpha_{u,s}(\lambda)\,ds$$

has the unique solution $\alpha_{u,t}(\lambda) = \exp(-\lambda^2(t-u)/2)$ a.e. $\widetilde{P}$. This means that X is a Brownian motion on $(\Omega, \mathbf{B}, P)$. □

There is a well-known sufficient condition for $E(M_{0,T})$ to equal 1. The condition is that $|f(s, \omega)| \le M$ ($f \in \mathcal{F}$ is bounded). A better sufficient condition has been given by A. A. Novikov (1973).

Historical Notes

Chapter 1. A. N. Kolmogorov (1933) was the first to establish a mathematical foundation of the theory of probability.

§1.1–§1.4. The results and their proofs are given in introductory books on probability theory. For example, A. N. Shiryaev (1979) discusses all the results in §1.1–1.4. The convergence theorems in §1.4 are also presented in Chapter 1 of the book by T. Hida (1980), where the description is well adapted to the present book. Although convergence in law is regarded as the weakest among the various types of convergence for a sequence of random variables, we know that it is not so far from almost sure convergence, in a certain sense. For this fact we refer to e.g. A. V. Skorokhod (1961) (English translation, 1965) Chapter 1, §6: If a sequence $\{\Phi_n ;\ n = 1, 2, \ldots\}$ of d-dimensional distributions converges to Φ in law, there exists a probability space $(\Omega, \mathbf{B}, P)$ on which we can form a sequence of d-dimensional random vectors $\{X_n ;\ n = 1, 2, \ldots\}$ such that

(1) the probability distribution of X_n is equal to Φ_n for each n,
(2) X_n converges, almost surely, to a random vector X, the distribution of which is Φ.

Theorems §1.4 can also be found in the above mentioned book by Hida. The proof of Theorem 1.7 was first given by Kolmogorov (1933), where the readers can see how his method is useful. Regarding the limit theorems, the law of iterated logarithms by A. Y. Khinchine (1933) or by W. Feller (1943) is worth mentioning, although it does not appear in the present book.

The method used to define conditional expectation and probability in §1.5 was originally proposed by J. L. Doob (1953). Also the concept of a martingale was introduced in his book and developed, in particular, the fundamental inequality (Theorem 1.10) is proved there.

The well-known sufficient condition for sample paths of a given stochastic process to be continuous (Theorem 1.14) is due to Kolmogorov and Prokhorov (1956). The proof is found in many books e.g. A. V. Skorokhod (1961). Yu. V. Prokhorov discussed, in his paper (1956), a general theory of convergence in law on the space C of continuous functions. This leads to the sufficient condition for continuity mentioned above.

Chapter 2. §2.1, contains a quick review of a Gaussian system. Further details of the theory can be found in the books by Yu. A. Rozanov (1963), T. Hida (1975), and others.

For more details about results in §2.2, consult Chapter 1 of the book by Hida (1975), where the proof of Theorem 2.3 is given. Also, a method of complexifying a Gaussian system is discussed in Chapter 6 of the book mentioned above and the contents of §§2.4–5 are similar to the first quarter of that book, although the description here is somewhat modified. As for Brownian motion, we give only the short background necessary for the theory of the canonical representations. Further information can be found in Chapter 2 of Hida (1975).

The definition of the Wiener integral was originally given by N. Wiener, as one can guess from its name. His idea can be seen in his book (1958), where he started with the integral by taking the integrand φ to be a C^1-class function with compact support. Namely, the integral is defined in such a way that

$$\int_0^\infty \varphi(t)\,dB(t) = -\int_0^\infty \varphi'(t)B(t)\,dt,$$

and then it was extended to a general case where φ is an element of $L^2([0, \infty))$. This definition is equivalent to that given in §2.5.

The evaluation for the modulus of continuity of Brownian paths was obtained in the best possible form by T. Sirao (1960). We state, in this book, continuity properties that are not of this form, but in a form (Theorem 2.9) sufficient for our purpose.

Chapter 3. The basic properties of stationary Gaussian processes are well known; we refer e.g. to I. A. Ibragimov and Yu. A. Rozanov (1970). The Wold decomposition of a stationary process is illustrated in H. Wold (1938), and its spectral decomposition method, used in §3.1, is available in M. H. Stone (1932) and K. Yosida (1965). The notion of the Hardy class $\mathbf{H}_2$, which we use in §§3.3–4, is explained in the book by K. Hoffman (1962). The moving average representation of a stationary process with continuous parameter was established by K. Karhunen (1950), so the representation (3.36) is called the Karhunen representation. In the application of the theory to a stationary Gaussian process, we form the canonical representation. Further applications to statistical estimation theory and others theories have been developed by U. Grenander and M. Rosenblatt (1957).

Chapter 4. The idea of the canonical representation of a Gaussian process was first proposed by P. Lévy in 1955 at the third Berkeley Symposium on Mathematical Statistics and Probability. The general theory of the representation has been developed by T. Hida (1960) and H. Cramér (1961), where they follow Lévy's idea and use the theory of spectral multiplicity. The basic idea in Hida (1960) is the use of the Hellinger-Hahn theorem for the reproducing kernel Hilbert space constructed from the covariance function $\Gamma(s, t)$

of a Gaussian process in question. In the present book, a system of adapted martingales are directly formed from a Gaussian process X, and the innovation system B of (4.1) is constructed in §4.2, as suggested by the proof of the Hellinger-Hahn theorem. Then follows, in §4.3, a correspondence between the representation space and the reproducing kernel Hilbert space. A recommended reference for the reproducing kernel Hilbert spaces is N. Aronszajn (1950). As for the Hellinger-Harn theorem, the book by M. H. Stone (1932) gives well established information.

The criterion (Theorem 4.4) for a given representation to be canonical is an extension of the result in Hida (1960), where he gave the criterion for a process with unit multiplicity. A method for constructing examples of a Gaussian process with higher multiplicity is given in M. Hitsuda (1973) and L. Pitt (1975). Examples of a noncanonical representation are found in Lévy (1956). They illustrate the meaning behind the definition of the canonical property and give us further questions to answer.

U. Grenander and M. Rosenblatt (1957), and M. Rosenblatt (1985) describe the theory of estimations in a systematic way and we can see that there are connections with what we have obtained in Chapter 4.

Chapter 5. As discussed in Chapter 5, to understand the multiple Markov property and to give an acceptable definition are profound questions. Many attempts have been made in this direction. The first one may be the work by J. L. Doob (1944). Shortly after the work of Doob, P. Lévy discussed these questions in the remarkable lecture at the Third Berkeley Symposium in 1955. In his lecture, he took up the $M(t)$-process as a basic example to give new insight and proposed a method to study the multiple Markov property. A development in this line was made by Hida (1960). In addition, Lévy's idea is widely applied to Gaussian processes with a multi-dimensional parameter, in particular, Lévy's Brownian motion. H. P. McKean Jr. (1963) studied the Markov property of the Brownian motion. Furthermore, he obtained many important results on the Markov property of Gaussian random fields with N. Levinson (1964) and with H. Dym (1970). While Y. Okabe (1974) obtained beautiful results on stationary Gaussian processes by using analytic methods.

Once the Markov property is viewed in terms of propagation of randomness as the time changes, a generalization of this concept may be found by taking the change of time variable into account. One such generalization would be the concept of T-positivity. Motivated by quantum field theory, this is taken to be a new concept in probability theory. The framework of G. C. Hegerfeldt (1974) is employed in this book. For readers who wish to understand it further we recommend referring to the above paper and consulting the references there. Also see T. Hida and L. Streit (1977). In addtion, there is a survey paper by Y. Okabe (1981). Papers by A. Noda (1987) and Si Si (1987) include new results related to Lévy's Brownian motion and to canonical representation theory. The present authors are grateful to Professor H. Oodaira for comments on the draft of the paper by Hida and Streit and for suggesting formula (5.92) for $\gamma(t)$, and to Professor H. Watanabe who informed us of the result of S. Bernstein (1928).

Chapter 6. An equivalence problem for measures on infinite dimensional spaces appeared in the pioneering work of S. Kakutani (1948). This was extended to a general theory of equivalence of Gaussian measures as is seen in Chapter 1 of Hida (1975). We use the method of J. Hajek (1958) and J. Feldman (1958) to discuss this equivalence problem in §6.2. Yu. V. Rozanov gave a clear proof of this theorem in his paper (1968), but the proof can now be simplified by using martingales.

As for equivalence problems related to Brownian motion, there have been many papers following after R. Cameron and W. T. Martin (1944). A necessary and sufficient condition for a Gaussian process to be equivalent to a Brownian motion were given by Hitsuda (1968), on which the descriptions in §§6.3–4 depend, although the arguments here are simplified. Martingale theory is used heavily in §6.4, and an overview is provided in the Appendix.

For the theory of integral equations of Volterra type, the book by F. Smithesis (1958) is a good reference. A condition for the equivalence in terms of the covariance operator is found in L. A. Shepp (1966). The topics presented in §6.5 are generalized by M. Hitsuda and H. Watanabe (1976), where an equivalence problem for a Gaussian process with spectral measure having discrete component is discussed. The relation between an equivalence problem and an integral operator of Volterra type was also obtained by I. C. Gohberg and M. G. Krein (1967). An interesting discussion about this idea is in the paper by G. Kallianpur and H. Oodaira (1973).

The innovation theorem, Theorem 6.5 was given independently by A. N. Shiryaev (1966) and T. Kailath (1968). A wide range of applications of the theorem are given in A. N. Liptzer and A. N. Shiryaev (1974). Applications of the theorem to Gaussian channels are treated from a viewpoint of information theory, in S. Ihara (1974) and Hitsuda and Ihara (1975).

Appendix. There are many ways to define a multiple Wiener integral. N. Wiener (1938) introduced the concept, which he called a homogeneous chaos. K. Itô (1951) defined a multiple Wiener integral by taking the symmetric L^2-kernel. Itô's definition is, of course, equivalent to the one given in §A.1, which starts from the integral of Volterra kernels.

Many years back, R. Cameron (and W. T. Martin (1947) gave the direct sum decomposition $L^2(\Omega, \mathbf{B}(B), P) = \sum_{i=0}^{\infty} \oplus \mathscr{H}_i$ in terms of the $\mathscr{H}_i$ which is in agreement with the space of multiple Wiener integrals of degree i. There the Hermite polynomials play an important role.

Several concepts taken up so far are intimately related in a profound level. The reason why we used Volterra kernels, as did Itô, is that the diagonal part of the kernel can be ignored so that the orthogonalization is done automatically. This trick may be rephrased as the Wick ordering in quantum dynamics.

Results on a martingale with continuous parameter came from the original work by J. L. Doob (1953). The stochastic integral dates back to K. Itô (1944), but at present there are many references on this subject, for example, H. P. McKean, Jr. (1969), S. Watanabe (1975), and so forth. Theorem A.9 was proved by H. Kunita and S. Watanabe (1967) as a corollary to the representation theorem of square integrable martingales.

The expression of a positive martingale on the Wiener space appeared in the paper by G. Maruyama (1954), where an equivalence problem for diffusion processes was first formulated. Theorem A.11 can be found in I. V. Girsanov (1960). Sufficient conditions for $E(M_{0,T}) = 1$ are presented in the book by R. Sh. Liptzer and A. N. Shiryaev (1974), which includes the well-known result of A. A. Novikov (1972).

References

N. Aronszain (1950), *Theory of reproducing kernels*, Trans. Amer. Math. Soc. **68**, 337–404.

S. Bernstein (1928), *Sur les fonctions absolument monotones*, Acta Math. **52**, 1–66.

——— (1938), *Equations différentielles stochastiques*, Actualités Math. **738**, 5–32.

R. Cameron and W. T. Martin (1944), *Transformation of Wiener integrals under translation*, Ann. of Math. (2) **45**, 386–396.

——— (1947), *The orthogonal development of non-linear functionals in series of Fourier-Hermite functionals*, Ann. of Math. (2) **48**, 385–392.

H. Cramér (1961), *On some class of non-stationary stochastic processes*, Proc. 4th Berkeley Symp. on Math. Statist. Probab. 2, Univ. Calif. Press, Berkeley and Los Angeles, CA, 57–78.

J. L. Doob (1944), *The elementary Gaussian processes*, Ann. of Math. Statist. **15**, 229–282.

——— (1953), *Stochastic processes*, Wiley, New York.

H. Dym and H. P. McKean (1976), *Gaussian processes, function theory, and the inverse spectral problem*, Academic Press, San Diego, CA.

A. Einstein (1905), *Über die von der molekularkinetischen Theorie der Wärme geforderte Bewegung von in ruhenden Flüssigkeiten suspendierten Teilchen*, Ann. Physik. **17**, 549–560.

A. Ephremides and J. B. Thomes (eds.) (1973), *Random processes. Multiplicity theory and canonical decompositions*, Dawden, Hutchison and Ross. Inc., Stroudsburg, PA.

J. Feldman (1958), *Equivalence and perpendicularity of Gaussian processes*, Pacific J. Math. **8**, 699–708.

W. Feller (1943), *The general form of the so-called law of iterated logarithm*, Trans. Amer. Math. Soc. **54**, 373–402.

——— (1950, 3rd ed. 1968), *An introduction to probability theory and its applications*, vol. 1, Wiley, New York.

——— (1966, 2nd ed. 1971), *An introduction to probability theory and its applications*, vol. 2, Wiley, New York.

C. F. Gauss (1821), *Theoria combinationis observationum erroribus minimis obnoxiae*, Göttingen.

I. V. Girsanov (1960), *On transforming a certain class of stochastic process by absolutely continuous substitution of measures*, Theory Probab. Appl. **5**, 285–330. (Russian original 314–330).

B. V. Gnedenko (1960), *Über die Arbeiten von C. F. Gauss zur Wahrscheinlichkeitsrechnung, Leben und Werke*, Haude & Spererschе Verlagsbuchhandrung, Leipzig, pp. 193–204.

I. C. Gohberg and M. G. Krein (1967, English ed. 1970), *Theory and application on Volterra operators in Hilbert space*, Amer. Math. Soc., Providence, RI.

U. Grenander and M. Rosenblatt (1957), *Statistical analysis of stationary time series*, Wiley, New York.

J. Hajek (1958), *On a property of normal distributions of an arbitrary stochastic process*, Czechoslavak Math. J. **8**, 610–618. (Russian)

G. C. Hegerfeldt (1974), *From Euclidean to relativistic field and on the notion of Markoff fields*, Comm. Math. Phys. **35**, 155–171.

T. Hida (1960), *Canonical representations of Gaussian processes and their applications*, Mem. College Sci. Univ. Kyoto A. Math. **33**, 109–155.

——— (1975), *Brownian motion*, Iwanami Schoten, Tokyo. (Japanese)

——— (1980), *Brownian motion*, Springer-Verlag, Berlin and New York.

T. Hida and L. Streit (1977), *On quantum theory in terms of white noise*, Nagoya Math. J. **68**, 21–34.

M. Hitsuda (1968), *Representation of Gaussian processes equivalent to Wiener process*, Osaka J. Math. **5**, 299–312.

——— (1972), *Formula for Brownian partial derivatives*, Second Japan-USSR Sympos. on Probability Theory (Kyoto) 2, 111–114.

——— (1973), *Multiplicity of some class of Gaussian processes*, Nagoya Math. **52**, 39–46.

M. Hitsuda and S. Ihara (1975), *Gaussian channels and the optimal coding*, J. Multivariate Anal. **5**, 106–118.

M. Hitsuda and H. Watanabe (1977), *On a causal and causally invertible representation of equivalent Gaussian processes*, J. Multivariate Anal. **4**, 247–265.

K. Hoffman (1962), *Banach spaces of analytic functions*, Prentice Hall, Englewood Cliffs, NJ.

I. A. Ibragimov and Yu. A. Rozanov (1970), *Gaussian random processes*, "Nauka", Moscow. (Russian)

——— (1978), *Gaussian random processes*, Springer-Verlag, Berlin and New York.

S. Ihara, *Information theory*, World Sci. Publ., Singapore and Teaneck, NJ (to appear).

E. L. Ince (1926), *Ordinary differential equations*, Dover, New York.

K. Itô (1944), *Stochastic integral*, Proc. Imp. Acad. Tokyo **20**, 519–524.

——— (1951), *Multiple Wiener integral*, J. Math. Soc. Japan **3**, 157–169.

T. Kailath (1968), *An innovation approach to least-squares estimation*, Part 1, *Linear filtering in additive white noise*, IEEE Trans. Automat. Control, AC-**13**, 646–655.

S. Kakutani (1948), *On equivalence of infinite product measures*, Ann. of Math. **49**, 214–224.

G. Kallianpur and H. Oodaira (1973), *Non-anticipative representations of equivalent Gaussian processes*, Ann. Probab. **1**, 104–122.

K. Karhunen (1950), *Über die Struktur stationärer zufälliger Funktionen*, Ark. Mat. **1** no. 3, 141–160.

A. Kintchine (1933), *Asymptotische Gesetze der Wahrscheinlichkeitsrechnung*, Ergeb. Math. Grenzgeb., vol. 114, Springer-Verlag, Berlin and New York.

A. N. Kolmogorov (1933), *Grundbegriffe der Warscheilichkeitsrechnung*, Ergeb. Math. Grenzgeb., vol. 2, no. 3, Springer-Verlag, Berlin and New York.

H. Kunita and S. Watanabe (1967), *On square integrable martingales*, Nagoya Math. J. **30**, 209–245.

H.-H. Kuo (1975), *Gaussian measures in Banach Spaces*, Lecture Notes in Math., vol. 463, Springer-Verlag, Berlin and New York.

N. Levinson and H. P. McKean Jr. (1964), *Weighted trigonometrical approximation on R^1 with application to the germ field of a stationary Gaussian noise*, Acta Math. **112**, 99–143.

P. Lévy (1948, Second ed. 1965), *Processus stochastique et mouvement brownien*, Gauthier-Villars, Paris.

——— (1951), *Problèms concrets d'analyse fonctionnelle*, Gauthier-Villars, Paris.

——— (1953), *Random functions; general theory with special reference to Laplacian random functions*, Univ. Cal. Pub. Stat. **1**, 331–390.

——— (1956), *A special problem of Brownian motion and a general theory of Gaussian random functions*, Proc. 3rd Berkeley Sympos. on Math. Statist. Probab., vol. 2, Univ. California Press, Berkeley and Los Angeles, CA, pp. 133–175.

——— (1957), *Fonctions aleatoires a correlation lineare*, Illinois J. Math. **1**, 217–158.

——— (1980), *Collected Works*, vol. IV (D. Dugué ed.), Gauthier-Villars, Paris..

R. Sh. Liptzer and A. N. Shiryaev (1974), *Statistics of stochastic processes*, "Nauka", Moscow. (Russian)

——— (1977–1978), *Statistics of stochastic processes*, Springer-Verlag, Berlin and New York.

V. Mandrekar (1974), *On the multiple Markov property of Lévy-Hida for Gaussian processes*, Nagoya Math. J. **54**, 69–78.

G. Maruyama (1954), *On the transition probability functions of the Markov process*, Natural Sci. Report, Ochanomizu Univ. **5**, 10–20.

H. P. McKean Jr. (1963), *Brownian motion with several dimensional time*, Theory Probab. Appl. **8**, 335–354.
——— (1969), *Stochastic integrals*, Academic Press, San Diego, CA.
A. Noda (1987), *Generalized Radon transform and Lévy's Brownian motion*. I, Nagoya Math. J. **105**, 71–87; II, Nagoya Math. J. **105**, 89–107.
A. A. Novikov (1972), *On an equality for stochastic integrals*, Theory Probab. Appl. **17**, 717–720.
Y. Okabe (1974), *On the structure of splitting fields of stationary Gaussian processes with finite multiple Markovian property*, Nagoya Math. J. **54**, 191–213.
——— (1979), *On the stationary Gaussian process with T-positivity and its associated Langevin equation and S-matrix*, J. Fac. Sci. Univ. Tokyo Sect. IA Math. **26** no. 1, 115–165.
——— (1981), *On a stochastic differential equation for a stationary Gaussian process with T-positivity and the fluctuation-dissipation theorem*, J. Fac. Sci. Univ. Tokyo Sec. IA Math. **28** no. 2, 169–213.
L. D. Pitt (1975), *Hida-Cramér multiplicity theory for multiple Markov processes and Goursat representations*, Nagoya Math. J. **57**, 199–228.
Yu. V. Prokhorov (1956), *Convergence of random processes and limit theorems in probability theory*, Theory Probab. Appl. **1**, 157–214.
M. Rosenblatt (1985), *Stationary sequences and random fields*, Birkhäuser, Basel and Boston, MA.
Yu. A. Rozanov (1963, English transl., 1967), *Stationary random processes*, Holden-Day, San Francisco, CA.
——— (1968, English transl. 1971), *Infinite dimensional Gaussian distributions*, Proc. Steklov Inst. Math. **108**.
L. A. Shepp (1966), *Radon-Nikodym derivatives of Gaussian measures*, Ann. Math. Statist. **37**, 321–354.
A. N. Shiryaev (1966), *Stochastic equations for non-linear filterations related to jump-type Markov processes*, Problems Inform. Transmission **2**, 3–22. (Russian).
——— (1979), *Probability*, "Nauka", Moscow. (Russian)
——— (1984), *Probability*, Springer-Verlag, Berlin and New York.
B. Simon (1974), *The $P(\phi)^2$ Euclidean (Quantum) field theory*, Princeton Univ. Press, Princeton, NJ.
——— (1979), *Functional integration and quantum physics*, Academic Press, San Diego, CA.
T. Sirao (1960), *On the continuity of Brownian motion with a multidimensional parameter*, Nagoya Math. J. **16**, 135–156.
Si. Si (1987), *A note on Lévy's Brownian motion*, Nagoya Math. J. **108**, 121–130.
——— (1989), *A note on Lévy's Brownian motion*. II, Nagoya Math. J. **114**, 165–172.
——— (1991), *Innovation approach to Lévy's Brownian motion*, Proc. Preseminar for the International Conference on Gaussian Random Fields. Part 2, Nagoya, pp. 27–33.
A. V. Skorokhod A. V. (1961, English transl. 1965), *Studies in the theory of random processes*, Addison-Wesley, Reading, MA.
——— (1975), *On a generalization of a stochastic integral*, Theory Probab. Appl. **20**, 223–238.
F. Smithesis (1958), *Integral equations*, Cambridge Univ. Press, London and New York.
M. H. Stone (1932), *Linear transformations in Hilbert space and their applications to analysis*, Amer. Math. Soc., Providence, RI.
K. Urbanik (1962), *Generalized stationary processes of Markovian character*, Studia Math. **21**, 261–282.
K. L. Vaninskii and A. M. Yaglom (1990), *Stationary processes with a finite number of non-zero canonical correlation between future and past*, J. Time Ser. Anal. **11**, 361–375.
S. Watanabe (1975), *Stochastic differential equations*, Sangyo Toshyo Pub. (Japanese)
N. Wiener (1938), *The homogeneous chaos*, Amer. J. Math. **60**, 897–936.
——— (1958), *Nonlinear problems in random theory*, Technology Press of MIT Cambridge, MA. and Wiley, New York.
H. Wold (1938), *A study in the analysis of stationary time series*, Almqvist & Wiksell, Stockholm.
K. Yoshida (1965), *Functional analysis*, Springer-Verlag, Berlin and New York.

Subject Index

Recent Titles in This Series

(Continued from the front of this publication)

85 **Mikio Ise and Masaru Takeuchi,** Lie groups I, II, 1991
84 **Đáo Trọng Thi and A. T. Fomenko,** Minimal surfaces, stratified multivarifolds, and the Plateau problem, 1991
83 **N. I. Portenko,** Generalized diffusion processes, 1990
82 **Yasutaka Sibuya,** Linear differential equations in the complex domain: Problems of analytic continuation, 1990
81 **I. M. Gelfand and S. G. Gindikin, Editors,** Mathematical problems of tomography, 1990
80 **Junjiro Noguchi and Takushiro Ochiai,** Geometric function theory in several complex variables, 1990
79 **N. I. Akhiezer,** Elements of the theory of elliptic functions, 1990
78 **A. V. Skorokhod,** Asymptotic methods of the theory of stochastic differential equations, 1989
77 **V. M. Filippov,** Variational principles for nonpotential operators, 1989
76 **Phillip A. Griffiths,** Introduction to algebraic curves, 1989
75 **B. S. Kashin and A. A. Saakyan,** Orthogonal series, 1989
74 **V. I. Yudovich,** The linearization method in hydrodynamical stability theory, 1989
73 **Yu. G. Reshetnyak,** Space mappings with bounded distortion, 1989
72 **A. V. Pogorelev,** Bendings of surfaces and stability of shells, 1988
71 **A. S. Markus,** Introduction to the spectral theory of polynomial operator pencils, 1988
70 **N. I. Akhiezer,** Lectures on integral transforms, 1988
69 **V. N. Saliĭ,** Lattices with unique complements, 1988
68 **A. G. Postnikov,** Introduction to analytic number theory, 1988
67 **A. G. Dragalin,** Mathematical intuitionism: Introduction to proof theory, 1988
66 **Ye Yan-Qian,** Theory of limit cycles, 1986
65 **V. M. Zolotarev,** One-dimensional stable distributions, 1986
64 **M. M. Lavrent′ev, V. G. Romanov, and S. P. Shishat·skiĭ,** Ill-posed problems of mathematical physics and analysis, 1986
63 **Yu. M. Berezanskiĭ,** Selfadjoint operators in spaces of functions of infinitely many variables, 1986
62 **S. L. Krushkal′, B. N. Apanasov, and N. A. Gusevskiĭ,** Kleinian groups and uniformization in examples and problems, 1986
61 **B. V. Shabat,** Distribution of values of holomorphic mappings, 1985
60 **B. A. Kushner,** Lectures on constructive mathematical analysis, 1984
59 **G. P. Egorychev,** Integral representation and the computation of combinatorial sums, 1984
58 **L. A. Aĭzenberg and A. P. Yuzhakov,** Integral representations and residues in multidimensional complex analysis, 1983
57 **V. N. Monakhov,** Boundary-value problems with free boundaries for elliptic systems of equations, 1983
56 **L. A. Aĭzenberg and Sh. A. Dautov,** Differential forms orthogonal to holomorphic functions or forms, and their properties, 1983
55 **B. L. Roždestvenskiĭ and N. N. Janenko,** Systems of quasilinear equations and their applications to gas dynamics, 1983
54 **S. G. Kreĭn, Ju. I. Petunin, and E. M. Semenov,** Interpolation of linear operators, 1982
53 **N. N. Čencov,** Statistical decision rules and optimal inference, 1981
52 **G. I. Èskin,** Boundary value problems for elliptic pseudodifferential equations, 1981

(See the AMS catalog for earlier titles)